# Issues in Cultural Tourism Studies

*Issues in Cultural Tourism Studies* examines the phenomenon of cultural tourism in its broadest sense. Drawing on postmodern perspectives, it emphasises the importance of popular cultural tourism; alternative and ethnic tourism; and working-class heritage and culture. Its main focus is on the role that cultural tourism plays in the globalisation process and the impacts of global tourism development on culture, traditions and identity, especially for regional, ethnic and minority groups.

The text outlines the development of cultural tourism, before discussing the environmental, socio-cultural and economic impacts that it can have on destinations and host communities. It then provides an analysis of the importance of cultural politics, especially for cultural tourism policy development. Later chapters discuss the role of cultural tourism in urban regeneration and also explore the organisational and policy framework for European cultural tourism. The text highlights the need for greater emphasis to be placed on the consideration of marginalised groups in policy formation for and in the management of cultural tourism. Individual chapters make specific references to the problems of exclusion and discrimination, as well as discussing issues relating to integration and identity.

*Issues in Cultural Tourism Studies* combines a rigorous and academic theoretical framework with practical case studies and real-life examples, initiatives and projects drawn from both the developed and developing world. It argues that the future development and management of cultural tourism relies on a greater degree of mutual understanding and communication between the sectors involved in its development, if it is to be sustainable, integrative and democratic.

**Melanie K. Smith** is Senior Lecturer in Cultural Tourism Management and Director of MA Programmes in Cultural Tourism/Cultural Industries/Heritage/Arts and Museum Management at the University of Greenwich.

# Issues in Cultural Tourism Studies

Melanie K. Smith

LONDON AND NEW YORK

First published 2003
by Routledge
11 New Fetter Lane, London EC4P 4EE

Simultaneously published in the USA and Canada
by Routledge
29 West 35th Street, New York, NY 10001

*Routledge is an imprint of the Taylor and Francis Group*



Typeset in Times by Keystroke, Jacaranda Lodge, Wolverhampton
Printed and bound in Great Britain by TJ International Ltd, Padstow, Cornwall



*British Library Cataloguing in Publication Data*
A catalogue record for this book is available from the British Library

*Library of Congress Cataloging in Publication Data*
A catalog record for this book has been requested

ISBN 0–415–25637–2 (hbk)
ISBN 0–415–25638–0 (pbk)

**I would like to dedicate this book to my wonderful Mum and Dad who instilled in me the great love of culture and travel that made this book possible.**

# Contents

# Preface

It would not be difficult to say what prompted the writing of this book, since much of the inspiration and motivation has been derived from years spent as an almost insatiable cultural tourist. This includes the various childhood day trips to castles, palaces and historic towns in the UK; numerous inter-rails 'doing' one European cultural destination a day on one crust of bread; the train-spotter-style ticking off of World Heritage Sites while backpacking around India and Southeast Asia; and the more recent long weekends in the culturally regenerated cities of Northern England. Not all exotic locations perhaps, but they aptly demonstrate the increasing diversity of the cultural tourism product and the activities that can be said to fall within its scope. Both tourism and culture have always been great passions of mine; hence it is fortuitous to have the opportunity to work within a field that allows me to indulge my passions, and to write about the interrelationship between the two phenomena. And, who knows, the royalties from this book might even pay for my next cultural trip!

On a more academic note, the rationale for writing this book is fairly clear. Five years ago, we developed an MA in Cultural Tourism Management at the University of Greenwich. At that time it was the first course of its kind in the UK, and perhaps even in Europe. Since that time, we have also infused our BA Tourism Management course with a cultural tourism flavour. This largely reflects staff research interests, but it is also in line with recent developments in the tourism and cultural industries.

The book aims to be contemporary in its focus, reflecting recent developments in the tourism and cultural industries, both in the UK and internationally. In recent years there have been many interesting examples of cultural tourism initiatives in the UK. For example, the British Tourist Authority has established a Cultural Tourism Group, the Scottish and Northern Ireland Tourist Boards have developed cultural tourism strategies, and many regional tourist boards are working closely with regional Arts Boards to promote arts and cultural tourism. It is gradually being accepted that audience development and the creation of revenue through tourism could help to ease funding crises within the arts, especially in regional or rural areas. Similarly, the development of cultural tourism can help to subsidise the expensive process of conserving heritage sites. Heritage management forms an integral part of sustainable tourism development, one of the areas which has been given priority by DCMS (the Department for Culture, Media and Sport). This is also clearly an international issue. Growing concerns for the environment and recognition of the need for integrated conservation strategies for global heritage have started to necessitate collaboration between tourism and heritage organisations at all levels, especially in the context of World Heritage Sites.

Cultural tourism has also been recognised as a growth area outside the UK. On a European scale, organisations such as ATLAS (the Association for Tourism and Leisure Education and Research), ECTARC (the European Centre for Traditional and Regional Cultures), the Council of Europe and the European Commission have been carrying out research in the field of cultural tourism and piloting initiatives since the late 1980s. It is also interesting to note that a number of the emergent Central and Eastern European destinations are focusing primarily on cultural tourism development.

Internationally, there has been a growing interest in the development of indigenous cultural tourism, which, if managed effectively, is often viewed as a means of revitalising or reinforcing the cultural traditions and identity of indigenous communities (this is true of many native communities in Australia, New Zealand, North America, and some parts of Asia and Central and South America). The recent development of BA degree programmes in cultural tourism in Australia, for example, tend to focus predominantly on the management of indigenous cultural tourism.

Like all forms of tourism, cultural tourism is currently a growth phenomenon; hence numerous destinations are developing different forms of cultural tourism throughout the world. This might include city-based tourism, the visiting of World Heritage Sites, festival attendance, or trekking in mountains, deserts or jungles, among other activities. The cultural tourism sector is potentially as diverse as any other; hence there is clearly a need for a more comprehensive analysis of the various facets of its development and management.

Cultural tourism studies is a relatively new and little known academic discipline, and one which might be described as a composite discipline, since it draws on a number of different academic areas for its theoretical underpinning. This includes areas such as anthropology, cultural studies, sociology, urban planning, arts management, heritage and museum studies, to name but a few. It could thus be argued that cultural tourism studies is one of the most fascinating and exciting new disciplines to emerge from the proliferation of tourism, leisure and cultural industries-based academic courses in recent years.

Cultural tourism education is consequently a new development; therefore it is unusual to see whole degree courses or training programmes devoted specifically to the subject. Instead, students tend to specialise in tourism, heritage, arts or leisure, choosing cultural tourism options where appropriate. However, there is evidence to suggest that there is a growing need for the potential workforce to be equipped with skills that enable them to work across all of the cultural sectors. It is arguably no longer enough to remain rooted firmly in one discipline or to specialise in one sector alone. The study of cultural tourism affords students and practitioners the opportunity to broaden their horizons and to gain a greater understanding of the interrelationships between the tourism and cultural sectors. There is also clearly a need for practitioners to move beyond a 'studies' approach to cultural tourism, and to consider the practical and vocational implications of cultural tourism development. Hence this book deals with many of the management issues that are pertinent to the cultural sectors.

It was evident to all of us while teaching cultural tourism and related courses that there are very few books devoted specifically to this phenomenon. Students are often required to consult or purchase several books in order to gain enough information to understand all the key issues and to undertake coursework. There was a definite need for a comprehensive core textbook for such students, but also for those practitioners with an interest in the interrelationships between tourism and the cultural sectors.

The writing of this book was therefore based on a perceived gap in the international market for a comprehensive book, which analyses the phenomenon of cultural tourism in its broadest sense. Existing texts (especially in Europe) have tended to concentrate on cultural tourism as a form of attractions or resources management, and the focus has often been on heritage tourism or the management of cultural events. Few writers have considered cultural tourism from a postmodern perspective; therefore emphasis has inevitably been placed on traditional or elite forms of culture.

The significance of popular culture, working-class, ethnic and gay culture has often been overlooked. This book will therefore aim to redress the balance by emphasising the importance of popular cultural tourism, of alternative or ethnic arts tourism, and of working-class heritage and culture, especially in industrial and agricultural areas. It will contrast some of these relatively recent areas of cultural tourism development with more established forms of cultural tourism, such as heritage tourism, indigenous tourism, urban tourism, and festival and events tourism.

Many writers in the field of indigenous cultural tourism have focused on the political nature of cultural tourism, where representation of culture, and the protection of cultural identity and traditions are central to its development and management. Much of this research has been carried out in post-colonial societies, where tourism is often viewed as a new form of imperialism, and the relationship between hosts and guests (indigenous communities and tourists) requires sensitive management. In contrast, much less has been written about the management of cultural tourism in post-imperial societies, where Eurocentric or ethnocentric approaches to the management of cultural tourism may still prevail. This book will examine some of the problems of interpretation and representation in the field of cultural tourism in a range of contexts, both in Europe and internationally.

Overall, it is hoped that this book will be of use both to students and practitioners alike. It should be used not only by those students specifically studying options in cultural tourism, but also those with an interest in all aspects of culture, whether it relates to tourism, heritage, museums, the arts, leisure or entertainment. There is arguably something in this book for everyone regardless of his or her cultural background or specific cultural interests. It is important for all of us working within the fields of tourism and culture to be as well informed as possible about the relationships between the cultural sectors, their interdependence and potential for collaboration. Only then can we move towards a better understanding of some of the more important social, political and ethical issues which ultimately affect us all.

*Melanie Smith*
*University of Greenwich*
*October 2002*

# Acknowledgements

I would like to thank a number of people for their support, help and encouragement during the writing of this book.

First of all, my friends at work (the girls and Ken) for their constant moral, emotional and practical support, particularly in recent times. Thanks also to numerous colleagues on the tourism conference circuit for their useful advice, encouragement, and for being such a source of inspiration to a young(ish) researcher.

Special thanks to Paul and Max for being my best friends and travelling companions for so many years, and to my parents and family for all of their love, support, encouragement and more. You always knew I wouldn't stay in the 'real world' for long!

# Introduction

Culture, like God and politics is everywhere.

(Adair, 1982: 12)

The reader might be forgiven for thinking that there is little need for yet another book about culture. Books and articles on this theme abound, and indeed are difficult to avoid. The post-colonial, postmodern era has afforded writers and researchers a wealth of cultural issues on which to cogitate and ruminate, including the globalisation phenomenon, which has brought numerous impacts in its wake.

However, the aim of this book is not to provide yet another overview of culture per se, but to discuss some of the issues pertaining to cultural development within the context of tourism studies. Cultural tourism studies is a composite discipline, hence one that necessitates an in-depth analysis of many relevant and contemporary social, political and ethical issues. This book does not, of course, claim to be entirely comprehensive in its scope, nor does it pretend to cover all of the complex theoretical arguments that underpin cultural development. What it does attempt to do, however, is to provide something of a stimulus for debate, and to act as a springboard for further discussion and research. The text provides a synthesis of many key ideas drawn from a diverse range of fields, and serves to encourage a more interdisciplinary approach to the study of cultural tourism.

Although the book should be thought provoking, it is not my aim to be overly contentious. The suggestion of a postmodern perspective may strike some readers as controversial, even alienating, but the intention was to adopt the most appropriate theoretical perspective for the subject matter. Hence postmodernism functions here largely as a prism through which we can view the multiplicity of perspectives of heterogeneous groups. A more detailed justification for this perspective is given in Chapter 1, which also analyses the profile of the so-called 'post-tourist'.

The theoretical framework for the book is derived partly from the field of cultural studies, and there are references to theorists from a range of other disciplines, including sociology, urban planning, arts, heritage and museum studies. The book is deliberately not over-theorised and dispenses with unwieldy jargon wherever possible so as to maximise its accessibility and appeal to a broad readership, including practitioners. Readers who find themselves wishing to explore any of the specific theoretical issues in more depth should avail themselves of some of the secondary data sources, which may provide a richer source of information. The theory in this book hence serves to provide a conceptual framework for the discussion of pertinent issues rather than being an end in itself. As stated by Milner (1994: 4), 'Culture has become a theoretical problem for us only because it is already socially problematic'.

Castro-Gomez (2001) distinguishes between traditional theory, which naturalises culture and focuses on concepts such as aesthetics and harmony, and critical theory which emphasises the socio-political or conflictive aspect of culture where culture is a site of contested meaning. It places culture in the context of political economy, including imperialism, post-colonialism, capitalism and globalisation. This book tends more towards critical theory, and several key themes pervade the text and provide the context for analysis.

These include the historical backdrop of imperialism and Eurocentrism, which have traditionally shaped the historiographies and political economies of many regions of the world. Following on from this is the theme of post-colonialism, which serves as a framework for the analysis of multiculturalism and identity construction. The socio-economic and geopolitical implications of the globalisation process are also discussed in some depth, especially insofar as they relate to commercialisation of culture and the impact of tourism. Inherent in all of these themes are the broad, complex and often insidious axes of power, which serve to control all forms of social, economic and cultural development. The reader will note that the book is not overtly political, except in its recognition of the importance and relevance of cultural politics to any discussion of cultural tourism. For example, there is relatively little discussion of contextualised policy-making or government intervention either for tourism or for culture, largely because it was my aim to focus instead on the international or global factors that shape cultural tourism development. However, the book is fairly unequivocal in its rejection of the totalising discourses and grand narratives that have shaped the globe. This includes the political structures that have served traditionally to assert the centrality and superiority of Europe; the uneven and unequal distribution of wealth and opportunity engendered by global capitalism; and the oppression or marginalisation of indigenous, ethnic or minority groups.

The work of the cultural theorist Raymond Williams provides many of the definitions and perspectives on culture that are used within this book. His inclusive and democratic views of culture and cultural development are particularly fitting for a text that focuses predominantly on culture as a democratic and plural concept. Williams' work focuses in particular on the social and political ramifications of the class system, championing the culture of the masses. His work aimed to democratise cultural policy, challenging the perceived elitism of Arts Councils, and the past tendency to focus on so-called 'high' or 'elite' culture in cultural policy-making. The concept of policy in cultural studies is not limited to the arts and public administration, but is also linked strongly to politics, power struggles and empowerment.

Within the field of cultural studies, identity construction and representation have become recurrent themes, with racial and ethnic issues gaining significance from the mid-1980s onwards. Central to these ongoing debates are postmodern theories of politics, power and ideology, which question hegemonic, Eurocentric and ethnocentric approaches to the representation of the culture of 'Others'. Along with Stuart Hall, a number of other theorists such as Gilroy, Mercer, West and Said have focused on issues of ethnicity, challenging the perceived preoccupation with national identity. Hall writes of 'a politics of representation', and claims that there is a need for a stronger assertion of self-identity in order to eliminate stereotypes and combat misrepresentation. Mercer writes of African, Asian and Caribbean diasporas, and the emergence of hybridised identities. Hall also focuses on diasporic and plural identities, arguing that there has been a shift from nationalism to ethnicity in the formation of identity. Other post-colonial theorists such as Bhabha, Spivak and Sarup address similar issues. Bhabha's work on the concept of hybridisation (whereby two cultures retain their own distinctive characteristics, but form something new through a kind of fusion) is of particular interest. Spivak brings a feminist perspective to the post-colonial debate about identity construction and representation. She focuses on the concept of 'marginality' in her analysis of the 'epistemic violence' which governs history, imperialism and colonial discourse. As well as his critique of post-colonial studies, Sarup's work on the relationship between race, ethnicity and 'nation-ness' is fascinating. Some of these ideas are discussed in more detail in Chapters 4 and 8.

The work of Edward Said has been particularly influential, as it examines the relationship between the Occident and the Orient and the hegemonic nature of European culture and power. The high cultural humanism of European rule was informed by scholarship and rational enquiry, hence providing adequate justification for the subordination of Orientals, and the oppressive binarism of 'us' and 'them' or 'self' and 'Other'. Bhabha has argued that this binary opposition between power and powerlessness is over-simplified, leaving little room for negotiation or resistance. It is perhaps worth reading his critique of Said's *Orientalism*, where he suggests that there is a certain ambivalence in the power relationship between coloniser and colonised, and that their identities sometimes become elided. However, there is little doubt that indigenous people were subordinated under colonial rule. By 1914 Europe controlled about 85 per cent of the Earth, functioning as a metropolitan centre ruling distant territories. Europeans appeared to believe that they had an obligation and a right to subjugate indigenous peoples, who were often depicted as barbaric, savage or primitive. Policies of assimilation or annihilation often had a devastating impact on native traditions, lifestyles and cultural practices. Sarup notes the impact that this had on indigenous identity:

> Imperialism . . . is an act of geographical violence through which virtually every space in the world is explored, charted and finally brought under control. For the natives, the history of their colonial servitude is inaugurated by this loss to an outsider of the local place, whose concrete geographical identity must thereafter be searched for and somehow restored.
>
> (Sarup, 1996: 150)

Eurocentrism has been a major informing principle in our construction of world historiographies, and is based on the notion of European superiority in terms of technology, social thought and cultural forms. Eurocentric thought embodies:

> The notion of a single, originary center, namely Europe, out of which everything superior emerges; the geopolitical division between the homogeneous West and a substantive, exotic East or, rather, Europe the center, and the rest of the world, the periphery.
>
> (Dirlik, 1999: 13)

Dirlik (1999) suggests that any radical critique of Eurocentrism must take into consideration the contemporary questions of globalism and post-colonialism, both of which are recurrent themes within this book. Eurocentrism is, of course, only one centrism that has historically encompassed the globe, relocating whole societies and communities in space and time, and transforming their historical, social and cultural trajectories. Mowitt (2001) suggests that multiculturalism is effectively taking the place of Eurocentrism as being the focal core of humanities research. Some might argue that the new panoptic centre of world experience is America, especially if global capitalism is viewed as a phenomenon that has emanated largely from the USA. The history of America does not, of course, begin with the arrival of white settlers; however, it is often defined in terms of a narrative of white discovery and settlement. The discourse of indigenous groups and Black and Hispanic immigrants needs to be given equal weighting in any construction or affirmation of American identity. Nevertheless, the assertion of white supremacy, the assimilation, annihilation or denigration of indigenous and immigrant cultures appears to have been a common Euro-American trait.

In their post-September 11th analysis, Sardar and Wynn Davies (2002) suggest that the omnipresence of America is inescapable. The rest of the world is exposed to its politics, foreign policy, media and cultural products, while Americans are often less exposed to foreign influence than any other nation on Earth. They use the metaphor of the ubiquitous hamburger (frequently a symbol of much hated global consumerism) to describe the packaging and presentation of America to the world, implying that even if you take out the bits you don't like, the influence is still pervasive. America is described as a hyper-power, which has the potential to lay waste to indigenous cultures, echoing Ritzer's (1993) view that American culture has acquired 'obscene power' in the process of replicating itself in the rest of the world. However, they suggest that this process is not simply a form of cultural imperialism akin to that imposed by European empires. American products have become so desirable that the so-called 'victims' are complicit in the spread of American culture:

> There are no competing powers because there is only one power, only one source of law and order. In such a natural order, it makes little sense to talk of Empire and American imperialism; indeed, such rhetoric and analysis are dangerously obsolete. Empires require colonies in which unwilling folks are forced into becoming subject people; imperialism entails a dominating metropolis trying to capture the markets and impose its rule on a distant country. Today, the globe is much more like an extension of American society, where – mostly – all too willing individuals and communities embrace American cultures and values.
>
> (Sardar and Wynn Davies, 2002: 65)

It is of course, misrepresentative to imply that America is an homogeneous nation; hence Sardar and Wynn Davies (2002) attempt to distinguish between America the political and economic entity, and the culturally diverse America of indigenous peoples and ethnic immigrants. However, despite living in a so-called democracy, the citizens themselves have relatively little political influence on their government's global decisions (but this is not, of course, unique to America).

The concept of globalisation is discussed in more depth in Chapter 1. Although the apparent 'Americanisation' of the world's cultures is clearly not synonymous with the notion of globalisation, the countries that consider themselves to be key players in the global arena are predominantly Western capitalist nations. In such countries, consumption is viewed as a means of gaining identity and prestige, and 'sign values' (i.e. socially constructed prestige values) become even more important than either 'use value' or 'exchange value'. Sarup (1996) suggests that the market offers customised identity-making tools through which one can construct a kind of 'DIY self'! Baudrillard suggests that individuals are trapped in a world of simulacra or 'hyper-reality' (perpetuated mainly by television and mass media) where the spectacle and the real are indistinguishable. Of course, the tourism industry is one in which myths, fantasies, 'hyper-real' and simulated activities can flourish.

Tourism is often described as the quintessential global industry, although it hardly needs to be pointed out that it is (again) a Western-dominated phenomenon in which the only role played by 'peripheral' countries is that of host rather than generator. Many theorists have described tourism as a new form of imperialism, a viewpoint that is discussed in more detail in Chapter 3. Sarup (1996: 127) conceptualises tourism (at least travel to less developed countries) in terms of 'self' and 'Other' and the notion of identity:

> Tourism is physical and metaphorical. Travel is a fascinating metaphor because it refers not to the fixed but to a journey, a crossing from the familiar centre to the exotic periphery. . . . Tourism is also a metaphor for the imposition of the Western gaze. There is enjoyment by the rich of the exotic difference of the Other and exploitation too. Travelling has also become an increasingly popular way of 'discovering one's identity'.

Sarup (1996) describes how identity construction is based on theories such as socialisation or role theory, ideology (e.g. state apparatus), discourse theory, and disciplines and technologies of the self. There is no such thing as a homogenous identity; all are multiple or hybridised in this postmodern, postcolonial, globalised world of fragmentation and disorientation. Clearly, the concept of identity is linked closely to issues of interpretation and representation, which are pivotal to discussions about heritage and arts tourism. Sarup describes how more traditional theories of identity construction were linked closely to the dynamics of class, gender and race. However, the complexities of psychological and sociological factors must also be taken into consideration. For example, the exploration of 'existential authenticity' which can be sought through travel (see Chapter 2) can be an important part of identity construction.

The concept of identity is a pervasive theme throughout the book. For example, Chapter 4 deals specifically with issues of integration and identity within the European context, analysing the role that cultural tourism initiatives can have in the fostering of integration and the complex interplay of European, national, regional and local identity construction. Chapter 7 considers the impacts that imperialism had on the lifestyles, traditions and identities of indigenous peoples, and the ways in which cultural tourism can sometimes help to revive traditions, renew cultural pride, and strengthen identity reconstruction following its often inevitable suppression under colonial rule. Chapter 8 looks at the ways in which ethnic groups express their culture and identity, and the complex interrelationship between tourism and the arts, where tourism can sometimes serve to compromise artistic integrity or quality, or impact upon the authenticity of performance. However, alternatively, it can help to revitalise artistic traditions and lead to new, exciting and liberating hybridised forms of creative self-expression.

These themes provide some of the theoretical and conceptual frameworks for each of the chapters, and are pervasive if not always explicitly stated.

Chapter 1 analyses the development of tourism from a postmodern perspective, focusing in particular on the experience of the so-called 'post-tourist'. A broad definition of culture as being about both 'a whole way of life' and 'the arts and learning' is adopted from the outset; hence the breaking down of the boundaries between art and life or culture and society provides the framework for an analysis of the postmodern de-differentiation of culture, tourism, leisure and lifestyle activities. It has been suggested that the post-tourist makes little differentiation between these activities, and often prefers the world of simulations, simulacra and 'hyper-reality'. The concept of authenticity has also become rather nebulous in the world of the post-tourist for whom such subjective attributions are largely irrelevant. The nature of the attractions that appear to characterise the postmodern era of tourism, and which are enjoyed by the post-tourist are discussed in some detail here.

The globalisation process, characterised by time–space compression, has afforded increasing numbers of people the opportunities to travel further afield. However, it is noted that globalisation is an uneven and unequal process; thus any debates relating to global tourism or culture are generally politicised, especially in this post-colonial era. Equality of opportunity, the relationship between 'self' and 'Other', and the concept of tourism as a new form of imperialism are touched on here, but are developed further in subsequent chapters. It is clear that the post-tourist is predominantly the child of Western, capitalist, consumerist societies, hence s(he) is not necessarily representative of the global tourism industry. Nor does s(he) characterise the average cultural tourist for whom priorities and motivations tend to be very different. Chapter 2 explores this relationship in more detail.

Chapter 2 provides a framework for a comprehensive analysis of cultural tourism issues. Clear definitions are provided, as is an analysis of both supply and demand in relation to the cultural tourism product. It is suggested that far from being a *niche* sector, cultural tourism is a growth sector, and one which increasingly needs to be differentiated in terms of activities, typologies and consumers. Although many cultural tourists tend to conform to typologies of tourism, which might categorise them as 'allocentric', 'explorers' or 'drifters' as definitions of cultural tourism broaden and diversify, it becomes more difficult to pigeonhole them into specific categories. It may become more appropriate to adopt more fluid typologies, or to consider them in terms of 'metensomatosis' (Seaton, 2002); that is, role-playing or the adoption of multiple personae. The latter part of the chapter outlines some of the generic issues that will be discussed in subsequent chapters. This list is by no means exhaustive, but it attempts to demonstrate the growing diversity of the cultural tourism industry and the issues that relate to it.

Chapter 3 focuses on the impacts of cultural tourism development in a range of contexts, placing emphasis on developing countries and traditional societies where the impacts of tourism are likely to be most significant. Some attention is paid to the debate about the extent to which tourism may be considered to be a new form of imperialism. Economically, tourism might appear to be the best development option for the future growth of a country or region, but the social, cultural and environmental consequences of such a decision must be considered carefully. The discussion therefore focuses on the management of socio-economic and socio-cultural impacts, exploring possible strategies for future planning and management. However, the frontiers of tourism are clearly being pushed to the limit; therefore few regions of the world are likely to escape its inexorable reach. As tourists search for more and more remote and isolated surroundings, encapsulated perhaps in the 'perfect' beach or trips into outer space, existing destinations struggle to cope with the legacy of tourism. At the same time, they may be attempting to retain visitor numbers, to upgrade or diversify their product, and to enhance their image. The balance is a difficult one, and although cultural tourism development can facilitate this process, it is not always the most appropriate development option.

Chapter 4 looks at the development of cultural tourism in Europe. Since the decolonisation of its empires, Europe has been grappling with issues of identity construction, both in terms of European Union and the nation state, but, perhaps more recently, that of the regions, many of which have asserted or are seeking political and cultural autonomy. Coupled with this apparently deep-seated identity crisis, many European countries are also faced with issues of integration of immigrants from former colonies, guest workers from former communist countries, or refugees forced from their homes by political crises. European society has become increasingly diverse, multicultural and multi-ethnic in this post-colonial era; hence there is a need to address these issues in cultural policy-making, and in the interpretation and representation of culture and heritage. Many trans-European initiatives have subsequently been developed in recent years to take account of the common yet diverse heritage and culture of Europe, and to help foster integration. Many countries are also diversifying their tourism industries and focusing increasingly on cultural tourism as a means of attracting visitors, not only to urban centres, but increasingly to remote and rural areas. This is especially true of emergent Central and Eastern European destinations.

Chapter 5 explores the relationship between cultural tourism development and issues of interpretation and representation, particularly within the heritage and museum sectors. The chapter first reviews the relationship between history and heritage, including some of the problems of depicting an accurate picture of the past. Some of the difficulties of interpreting more sensitive or dissonant heritage are also discussed in some detail. The commercialisation or commodification of heritage can sometimes be exacerbated by tourism development, especially if the authenticity of a site is compromised for the purposes of entertainment. Similarly, there are some concerns within the museum world that the demands of tourism are leading to the erosion of the traditional collections-based museum. The changing role of museums is considered in depth, including a discussion of the need for a more inclusive approach to interpretation and representation of collections and exhibitions.

Chapter 6 focuses on the globalisation of heritage tourism, in particular the development of the World Heritage List. Emphasis is also placed on some of the complex management issues that affect all heritage sites, but especially those of 'universal value'. The difficulties of maintaining a balance between the creation of access for visitors and the need for conservation are explored in some detail, as are problems of 'ownership' and multiple interpretations. The chapter discusses the concept of World Heritage Site status, its meaning and significance, and the way in which the inscription process is gradually adopting a less Eurocentric focus. Some of this debate will centre around the notion of 'aesthetics' which is gradually being discredited as a marker of universal value. Instead, 'historicity' is becoming a much more valuable and valued indicator of political, social and cultural significance, and one which is being used increasingly to designate and inscribe World Heritage Sites.

Chapter 7 examines the political, social and economic situation of indigenous peoples in a variety of contexts, and assesses the extent to which cultural tourism may be considered to be a positive development option. As well as discussing the impacts of indigenous cultural tourism, the chapter focuses on the need for more community-based cultural tourism initiatives. An overview is provided of the measures that need to be taken in order to support, encourage and empower indigenous people so that they can eventually own and manage their own tourism-related ventures and initiatives. The cultural representation of indigenous peoples is also discussed, as it is still not uncommon for ethnic and tribal groups to be depicted as the 'exotic Other' in exhibitions, postcards or tourist literature. Some attention is also paid to the concept of authenticity of experience, as cultural performances frequently provide an arena in which indigenous people again form the focus of the commodified tourist gaze.

Chapter 8 deals with the often fraught relationship between cultural tourism and the arts, examining some of the past tensions that have acted as a barrier to the development of arts tourism and collaborative initiatives. The chapter also deals with issues relating to access, democracy and inclusion, and the need for the arts to become more accessible to wider audiences, and to promote lesser-known arts activities. This includes many of the ethnic and minority artistic endeavours, which have traditionally been ignored or under-funded. Emphasis is placed in particular on festivals, events and carnivals, which are arguably more inclusive and participatory, as well as being largely community based. The impacts of tourism on such art forms is also considered in terms of commodification, authenticity and identity.

Chapter 9 looks at the phenomenon of urban regeneration and the role that cultural tourism can play in this process. This includes the development of cultural quarters within cities, the use of cultural 'flagships' as catalysts for further social and economic development, and the enhancement of external image through cultural tourism. Emphasis is placed mainly on European industrial cities whose traditional industries have declined; hence culture and tourism are being viewed increasingly as a means of compensating for decline or for revitalising the economy of such cities. However, it is recognised that many of these cities have followed the example of a number of American cities, which have been transformed through regeneration. Although cultural tourism is not viewed as a panacea for the redevelopment of such cities, it is certainly a stimulus for economic diversification.

However, the chapter asks questions about the way in which the arts and culture are being used increasingly as tools for economic regeneration, rather than being valued for their intrinsic qualities. The latter part of the chapter also looks at the way in which regeneration is helping to produce new spaces for tourism, leisure and recreation, at the same time as transforming and 're-creating' old places. Although this may be viewed as a positive process, care must be taken not to impose a globalised, bland standardisation of development. This chapter takes us back to our starting point in Chapter 1, which questioned the extent to which postmodern tourism revolves around the experience of difference, or the reassuring blanket of sameness, familiarity and de-differentiation. The conclusion must surely be that distinctive and distinguishing characteristics of place must be preserved (or created) if we are to sustain a tourism industry which embraces uniqueness and champions cultural diversity.

# 1 A framework for cultural tourism studies

> There comes to be generated a kind of stylistic melting-pot, of the old and the new, of the nostalgic and the futuristic, of the 'natural' and the 'artificial', of the youthful and the mature, of high culture and of low, and of modernism and the postmodern.
>
> (Urry, 1990: 90)

## Introduction

The aim of this Chapter is to provide a framework for the rest of the book in terms of definitions, contexts and perspectives. Whereas much of the book focuses on different typologies of cultural tourism and interrelationships between the cultural sectors, this chapter is more conceptual, introducing many of the recurrent themes that underpin the text. This includes the historical context of imperialism and post-colonialism that has given rise to the redefinition of cultural boundaries and identities; the geographical, social and political location of cultural tourism within the globalisation process; and an analysis of the validity of a postmodern approach to cultural tourism studies. As a multi-disciplinary subject, cultural tourism studies clearly necessitates an awareness and understanding of a wide range of international issues. By providing an historical, political, social and geographical framework for the analysis of cultural tourism, it is hoped that appreciation of the wide-reaching significance of this fascinating phenomenon will be furthered.

## Definitions of culture

The cultural theorist Raymond Williams (1976) once described 'culture' as one of the most complex words in the English language, and hence one of the most difficult to define. This comment notwithstanding, he went on to provide some of the most coherent and oft-quoted definitions of culture in the field of British cultural studies. Williams sought to democratise cultural theory. One of his central preoccupations was the perceived conservatism and elitism of cultural policy-making in the UK. He championed the cultural activities and interests of the masses, particularly working-class and rural communities, declaring that 'Culture is ordinary: that is the first fact. Every human society has its own shape, its own purposes, its own meanings. Every society expresses these, in institutions, and in arts and learning' (Williams, 1958: 4). He referred to culture as meaning 'a whole way of life – the common meanings', and 'the arts and learning – the special processes of discovery and creative effort' (ibid.: 8).

Williams clearly recognised the need for a convergence between both anthropological and sociological definitions of culture. Hence culture is viewed as being about the whole way of life of a particular people or social group with distinctive signifying systems involving all forms of social activity, and artistic or intellectual activities. These are useful and comprehensive definitions, since they cover both the development of individual and group culture, conveying the importance of heritage and tradition, as well as contemporary culture and lifestyles. Culture is not just about the arts and the aesthetic judgements of a select minority who have been educated to appreciate certain cultural activities; it is also about the lives and interests of ordinary people, both urban and rural dwellers, indigenous and immigrant communities, artists and artisans.

In many ways, Williams' notion of culture being a plural concept is derived from the ideas of earlier theorists such as Johann Gottfried Von Herder (1774), who perceived culture as being about distinctive ways of life rather than a universal value. He opposed the idea of a Eurocentric universal culture, which was often defined in relation to the colonial 'Others', stating that each people (society, ethnic group, linguistic community) could be distinguished by a 'whole way of life', by common customs, lifestyle, traditions. Herder questioned the distinctions of 'Zivilisation und Kultur', the counterpart tradition in Britain being the 'Culture and Society' or 'Culture and Civilisation' debates.

In the eighteenth and nineteenth centuries, the idea of culture was defined in opposition to nature as a range of social, political, ethical, religious, philosophical and technical values. A number of thinkers and philosophers such as Hegel deemed cultural forms that were close to nature as being inferior to those that were linked more closely to the human spirit. Hence, high culture is assumed superior to popular culture, and Western, predominantly Christian culture is considered superior to, for example, those of tribal peoples who lived closer to nature and often practised naturalistic rites. It was also believed by a number of key thinkers that the nation state was the true carrier of the culture or national spirit of a people (Castro-Gomez, 2001).

Cultural theorists who were part of the 'Culture and Civilisation' tradition, such as Matthew Arnold, emphasised the social and educational significance of Culture (with a capital 'C'), stating that: 'Culture seeks to do away with classes; to make the best that has been thought and known in the world current everywhere' (1875: 44). Arnold argued that culture brings enlightenment or 'cultivation', which transcends social divisions, such as class, gender, race, religion and ethnicity. It is interesting however to note that Arnold refers to 'Culture' rather than 'cultures', implying an aesthetically narrow, rather than a plural or diverse concept of culture. His theories are now also deemed essentially Eurocentric and elitist, especially given his condemnation of popular culture for lacking aesthetic value (Jordan and Weedon, 1995).

It has been recognised that there is a need for democratic and pluralist participation in the institutions and practices of culture. Recent cultural theorists tend to adopt a plural concept of culture, and recognise the diversity and hybridity of different cultures. For example, Hannerz (1990: 237) claims that everyone participates in many 'cultures':

> The word culture is created through the increasing interconnectedness of varied local cultures, as well as through the development of cultures without a clear anchorage in any one territory. These are all becoming sub-cultures, as it were, within the wider whole.

The world of global politics, media, communication and technology is a polarised one in which only a minority can truly participate. The majority of developing countries are afforded few opportunities to play their part in the global economy, except perhaps through tourism. Even then, this is likely to be only in terms of hosting rather than generating tourism. The colonial legacy of the Western world has led to a growth in multiculturalism and ethnic diversity, especially in urban environments, and the culture and identity of minority groups within the national context have become thc focus of numerous theoretical and political debates. Cultural protectionism, particularly in regional, rural and remote areas of the world, has arguably increased in response to the sweep of globalisation, as many indigenous and minority peoples feel the need to protect their interests in the face of this powerful, exclusive phenomenon. This can manifest itself in political struggles or can be expressed through the arts and culture.

Tourism has clearly become a global force dominated mainly by Western developed nations whose globe-trotting citizens have left few places unexplored. Only the remotest locations of the world are 'safe' from tourism, but even then, other global forces look set to encroach on such environments if tourism does not get there first. Tourism, and especially cultural tourism, has become a force to be reckoned with, irrevocably transforming destinations, traditions and lifestyles. For this reason, cultural tourism has become increasingly politicised as governments weigh up the advantages and disadvantages of this potentially lucrative industry, often viewing it as their sole economic option if they wish to compete in the global arena. The environmental and socio-cultural consequences of such decisions are often overlooked; hence it falls to the communities themselves to protect their own interests, usually with inadequate political support.

## The politics of cultural tourism

Eagleton (2000: 32) states that 'It is hard to resist the conclusion that the word "culture" is both too broad and too narrow to be greatly useful'. It is indeed difficult to establish parameters around definitions of culture in the postmodern, global world where culture could quite reasonably be defined as almost any activity that relates to the lives and lifestyles of human beings. Eagleton questions the all-encompassing definitions of culture which relate to lifestyles and cultural preferences; for example, such concepts as 'football culture', 'café culture' or 'museums culture'. Of course, such concepts are specific mainly to Western developed countries where culture is thought of frequently as a lifestyle choice, rather than an organic phenomenon pertaining to the whole way of life of a people. As stated by Butcher (2001b: 16):

> Culture, for those living in poorer countries, often corresponds more closely to the original meaning of the word – tending crops and animals – a way of life dictated by living on the margins of the global economy rather than a cultural lifestyle of choice.

Culture and commerce have clearly become intertwined in the context of the cultural and creative industries. In the postmodern world of global cultural consumption, culture has become a commodity to be packaged and sold much like any other. However, there are problems with the basic assumption that people and places can be treated similarly. For example, Hewison (1991: 175) argues that culture and commerce are not synonymous:

> The time has come to argue that commerce is not culture, whether we define culture as the pursuit of music, literature or the fine arts, or whether we adopt Raymond Williams's definition of culture as 'a whole way of life.' You cannot get a whole way of life into a Tesco's trolley or a V & A Enterprises shopping bag.

Some postmodern theory favours a more participatory and democratic approach to cultural development including the breaking down of barriers between culture and society, art and life, high and low culture. The process of globalisation has led to the hybridisation of different cultural forms, and decolonisation and immigration have contributed positively to the multiculturalism of the global city. However, issues pertaining to cultural integration, identity construction and representation have also been prominent in cultural studies literature over the past few decades.

The concept of culture has become increasingly politicised, especially where it is defined as a way of life of a people or society. Sarup (1996: 140) states that:

> Culture is not something fixed and frozen as the traditionalists would have us believe, but a process of constant struggle as cultures interact with each other and are affected by economic, political and social factors.

Within the context of British cultural studies, for example, Stuart Hall has argued that culture is inherently political. Eagleton also emphasises the strong political dimensions of culture in terms of both conflict and identity construction, particularly in multicultural, postcolonial societies:

> On a global scale, the relevant conflict here is between culture as commodity and culture as identity. The high culture of Bach and Proust can hardly compete as a material force with the seductions of the culture industry, a religious icon or a national flag.
>
> (Eagleton, 2000: 72)

Bradford *et al.* (2000: 339) state that:

> Culture, understood as the values, worldviews, and identities that people construct for themselves, plays a major role in world events. Culture affects the coherence and viability of nations. This is not the 'culture' of high society, the elite arts, or the commercial media. Rather it is the culture of ordinary people as expressed in daily life, on special occasions, and in trying times. Culture has emerged as a topic of public concern and political action.

If culture is politicised, so then is cultural tourism. As stated by Lanfant (1995: 4):

> tourism, particularly 'cultural tourism', is often considered by international organisations as a pedagogic instrument allowing new identities to emerge – identities corresponding to the new plural-ethnic or plural-state configurations which are forming.

She questions the extent to which the forming of identities is motivated by ideology. Tourism has frequently been described as an imperialist or hegemonic power (for example, see Chapter 3), and identity has become a product to be manufactured, packaged and marketed like another. However, Lanfant (1995: 5) suggests that it may be overly simplistic to describe tourism simply as a new form of imperialism, stating that:

> The tourist system of action is not a monolithic force. It would be pointless to seize upon it as if it were a hegemonic and imperialist power perpetuating disguised neo-colonialism. This system is a network of agents: these tap a variety of motivations which are difficult to define and which in concrete situations often contradict each other.

As Lanfant argues, most identities are constructed in relation to others. The philosophical basis of this argument may be derived from Existentialist theories of 'self' and 'other' whereby our sense of ourselves is partly defined by the 'other'. Edward Said clearly took this concept further within the context of colonial and post-colonial studies. Stuart Hall also explores such concepts in his work on the cultural politics of ethnicity. The concept of alterity has indeed become central to the discipline of cultural politics and cultural studies, and subsequently cultural tourism studies (e.g. see the work of Hollinshead, MacCannell, Nash, Selwyn, to name but a few). Both dominant and marginal groups within a host society are subjected to the 'gaze' of the tourist (Urry, 1990), often reducing them to inauthentic stereotypes. Their culture is somehow fossilised and romanticised under the tourist gaze. As stated by Kirschenblatt-Gimblett (1998: 54), the destination and its people can then become something of a museum or theme park:

> tourism more generally takes the spectator to the site, and as areas are canonized in a geography of attractions, whole territories become extended ethnographic theme parks. An ethnographic bell jar drops over the terrain. A neighbourhood, village, or region becomes for all intents and purposes a living museum in situ.

Without a strong sense of their own identity, host societies are likely to succumb to the temptations of commercialisation, and lose sight of their traditions. Although economic development and 'progress' should not be denied to such communities, care must be taken to ensure that the people themselves are in a position to make an informed decision about their destiny. The nature and scope of tourism development will, of course, be determined partly by the local natural and cultural resources available. Butcher (2001b: 15) suggests that:

> how the host is viewed through the prism of culture, inevitably affects the prospects for and type of development on offer. Culture defined as function and difference effectively creates culture as a straightjacket for societies that may desire economic development. Culture becomes objectified; a romantic image cast in stone, rather than the creative subjectivity of the host. It can become a part of heritage, the past, preserved for the sensibilities of the tourist, rather than being made and remade in the context of social change.

Culture is of course dynamic rather than static; hence this fossilisation of culture is arguably a misrepresentation of ongoing indigenous traditions (as discussed in more detail in Chapter 7). However, the Western obsession with heritage has often reduced indigenous culture to the status of a dead or lifeless phenomenon. The concept of the 'living museum' is described in detail by Kirschenblatt-Gimblett (1998: 151) who states that 'Heritage and tourism are collaborative industries, heritage converting locations into destinations and tourism making them economically viable as exhibits of themselves. Locations become museums of themselves within a tourism industry.' Heritage is, of course, a problematic and contentious concept, particularly in terms of interpretation and representation, which will be discussed in more detail in Chapter 5.

## Globalisation and cultural tourism development

As the world becomes increasingly globalised, some might argue that culture is becoming more standardised or homogeneous (after all, who can avoid Coca-Cola, Nike trainers and McDonald's, even in the furthest reaches of the globe?). The contemporary philosopher Simon May (1999: 84) stated that 'Fanfares for "cultural diversity" have come just when most of it has gone'. This is arguably a pessimist's view of the globalisation process, yet it has some important implications for our sense of cultural identity and proclaimed need for cultural protectionism. Yet it could also be argued that post-war decolonisation, mass immigration and the growth of international tourism have all contributed to an enrichment of the world's cultures, leading to a profusion of cultural diversity, and the creation of new, cosmopolitan cultural forms.

The world is apparently becoming a smaller place because of instantaneous transactions and communications, increased and faster travel, and technological innovations. English is becoming the global language, and culture is becoming more and more dominated by American or Western European models. The global economy tends to be led by a limited number of predominantly Western nations and corporations, who control and standardise economic and cultural production. There are therefore concerns that globalisation benefits only certain communities of the world and marginalises others, and it can have a significant impact on regional and local cultures, traditions and languages.

Robins (1997) describes globalisation as being about the dissolution of the old boundaries and structures of nation states and communities, and the increasing transnationalisation of economic and cultural life. He also sees globalisation as the growing mobility of goods, commodities, information, services and people across frontiers. Appadurai (1990) summarises these flows as:

- *Ethnoscapes* (e.g. flows of people, such as workers, immigrants, tourists).
- *Technoscapes* (e.g. flows of information, communications, technology).
- *Finanscapes* (e.g. flows of money, foreign exchange, financial transactions).
- *Mediascapes* (e.g. flows of images, satellite television, twenty-four-hour news).
- *Ideoscapes* (e.g. flows of ideas, such as politics, religion, environmentalism).

The world is characterised increasingly by international media, technology, communications, economics, transport and tourism. Time–space compression is largely a result of instantaneous transactions and transmissions, and the facilitation of global travel. Harvey (1990) discusses the concept of 'time–space' compression in some detail, especially in the context of the global city. He sees postmodernism partly as a response to a new set of experiences of space and time.

However, globalisation is clearly a complex and apparently contradictory phenomenon, which helps to create new spaces of commonality, but also new spaces of difference. As stated by Waters (1995: 136): 'Globalisation . . . is a differentiating as well as a homogenising process.' It may be viewed as a creative process which leads to the hybridisation of different cultural forms, but it can also be considered in terms of the homogenisation or standardisation of culture. Cultural hybridisation is arguably the exciting and liberating face of globalisation, whereby new cultural forms are created through the fusion of diverse elements. This is particularly evident in Western food, fashion and music, for example, where ethnic influences are becoming more and more pervasive. As stated by Beynon and Dunkerley (2000: 26):

> Ethnicity no longer resides in the narrowly local, as is witnessed in the proliferation of ethnic cuisine, ethnic fashion, ethnic holidays and ethnic music. All over the globe there has been an indigenisation of music, art, architecture, film and food and what was feared by many (namely Western cultural domination) is becoming less likely.

Urry (2002: 161) describes how the tourist gaze has now been globalised:

> We can talk of the globalising of the tourist gaze, as multiple gazes have become core to global cultures sweeping up almost everywhere in their awesome wake.

Tourism is a global phenomenon and there are few remaining uncharted territories, as will be discussed in Chapter 3. However, the globalisation process is clearly an uneven and unequal one; as stated by Li (2000), globalisation is not truly global. The monopolisation of the global economy and marketplace by multinational corporations has arguably led to a degree of manipulation in terms of the production and consumption of culture. This is an argument that is attributed largely to Adorno and Horkheimer, whose views could still be said to have some contemporary relevance.

## The development of the cultural and creative industries

The idea of a 'cultural industry' was first used by Adorno and Horkheimer (1947) who argued against the commonly held belief that the arts were independent of industry and commerce. They claimed that cultural items were produced in a similar way to any other consumer goods. They linked the idea of the 'cultural industry' to mass culture, arguing that the products were of a standardised or homogeneous nature, and that artistic integrity had been compromised by production methods. They drew on Marx, arguing that the nature of cultural production within a capitalist industry results in a standardised commercial, mass-produced commodity. Adorno and Horkheimer believed that individuals and groups were manipulated into pursuing certain activities by capitalist corporations and governments. Culture was deemed unchallenging and pacifying, lulling consumers into becoming unthinking 'masses'. Rockwell (1999) suggests that the Frankfurt School (of which Adorno and Horkheimer were a part) may have reacted strongly to seeing the Nazis and Stalinists crush artistic freedom in the name of the masses, hence likening the manipulation to a form of subtle totalitarianism.

They also considered that the separation of work and leisure was largely illusory, as many of the activities that people practised in their so-called spare time was as repetitive as work on the assembly line (jazz music and film are cited as two of the main culprits here). Production masquerades as freedom and escapism, but in fact draws the consumer deeper into a mire of drudgery. They also deemed that the cultural industries served to reduce or eradicate the distance between art and life, lulling audiences into passive consumption and predictable responses, rather than shocking or provoking them (it is worth noting that dramatists such as Brecht had addressed this notion of bourgeois complacency in his use of the 'Verfremdungs' or 'Alienation' effect). There is also the notion that people no longer responded spontaneously to works of art, but evaluated them only according to their value in the marketplace. Hence exchange-value largely replaces use-value, a theory that is derived from Marx's concept of commodity fetishism. People become enslaved by, rather than emancipated by art and culture, surrendering their autonomy.

Adorno reflected later in his essay of 1967 'Culture industry reconsidered' (see Adorno, 2001), on why he and Horkheimer had replaced the concept of 'mass culture' with 'cultural industry'. They believed that the products of mass culture did not emanate from the people themselves, but were administered by a central hegemonic authority. The Italian neo-Marxist Gramsci argued in his theory of hegemony that cultural producers were able, through various means, to win continually the consent of the masses.

MacDonald (1994: 30) states that 'The Lords of *kitsch*, in short, exploit the cultural needs of the masses in order to make a profit and/or to maintain their class rule'.

In 1982, UNESCO published *The Culture Industries: A Challenge for the Future of Culture*, which took Adorno and Horkheimer's essay as a starting point. They pluralised the concept of 'the cultural industries', re-emphasising the fact that consumer demand was dictated largely by powerful corporations, and that artistic creativity and spontaneity were being subordinated. UNESCO were also concerned at the way in which the entertainment industry was contributing to global inequality, undermining cultural diversity, and standardising or homogenising culture (Negus, 1997). The threat of increasing globalisation and deregulation had perhaps fuelled these concerns. Hewison (1987) viewed the heritage industry in much the same way as Adorno and Horkheimer's viewed the cultural industry: as an artificial history imposed on the public by marketing managers from above. Nevertheless, he supports the view that the public are largely manipulated in their choice and consumption of culture.

However, it could be argued that this critique of mass culture, homogenisation and the undifferentiated responses of the masses is in many ways elitist. Clearly, people's needs and responses vary enormously, and the majority of people are arguably informed enough to select their own cultural preferences. Popular or mass culture can break through traditional and hegemonic structures. MacDonald (1994: 32) states that 'Mass culture is a dynamic, revolutionary force, breaking down the old barriers of class, tradition, taste, and dissolving all cultural distinctions.' Postmodern theory would support this particular viewpoint.

## A postmodern perspective on cultural tourism studies

Postmodern theory is useful in the context of the previous debate, as it favours de-differentiation of cultural activities and preferences. However, the concept of postmodernism is often regarded with incomprehension or distrust. As stated by Adair (1982: 12), 'Few "isms" . . . have provoked as much perplexity and suspicion as postmodernism.' Solomon and Higgins (1996: 300) criticise postmodernism for its failure to contribute anything positive or new to philosophical discussion, stating that:

> Postmodernism . . . has come to represent a ragbag of objections, accusations, parodies, and satires on traditional philosophical concerns and pretensions. It is largely negative, rarely positive, the celebration of an ending but not clearly marking anything new.

Sarup (1996: 102) notes that 'many thinkers believe that the rhetoric of postmodernism is dangerous, for it avoids the reality of political economy and the circumstances of global power'. Walsh (1992) describes postmodernity as a condition, rather than as a coherent set of beliefs and practices. Solomon and Higgins (1996) argue that postmodernism is not a philosophy, since it rejects the concept of a single universal or absolute truth. Instead, its proponents argue in favour of plural and objective concepts of 'culture' and 'discourse'. Postmodern structuralists such as Levi-Strauss and Foucault have argued against universal and totalising discourses, the authority of the single subject, and the generally accepted linearity of history. Foucault in particular emphasises the power relationships and manipulation inherent in social structures, which affect cultural development.

Therefore, although postmodernism could be disregarded as just another contemporary 'buzzword', its influence is pervasive in cultural theory. However, Milner (1994: 137) describes it as 'a particular cultural space available for analysis to many different kinds of contemporary cultural politics'. Urry (1990) describes postmodernism as being 'anti-hierarchical' and opposed to differentiations. Modernists considered there to be a definite gap between high art and low art, whereas postmodernism questions these divisions. This shift in perspective has been essential to the development of cultural democracy, increased access to culture and inclusive representation. Walsh (1992: 54) describes how 'media of mass communications have facilitated the removal of many of the boundaries between high art and low art and have helped to remove difference from the varied and rich cultures all over the world', and that

> De-differentiation manifests itself in a number of ways. There is the destruction of the division between high and low art, the end of auratic, or rather, the end of the provision of auratic spectacles solely for consumption by a social elite. The de-differentiation of culture also results in the incorporation of culture into the everyday political economy.

Postmodernism is interesting in that it conceptualises culture less in terms of homogeneity, and more in terms of diversity, hybridisation and local discourses. This is particularly important in the increasingly cosmopolitan and culturally diverse environment of the global city. Knox (1993: 17) refers to the usefulness of postmodernism in an urban context: 'Postmodernity is an increasingly important dimension of socio-cultural life that articulates with certain features of economic and social change in contributing to the socio-spatial dialectics of the city.'

The concept of postmodernism seemed to crystallise in the mid-1970s in America, Britain and Europe, especially with the publication of Lyotard's *La Condition Postmoderne* in 1979. Strinati (1994: 429) analyses the distinguishing features that characterise postmodernism as:

1. The breakdown of the distinction between culture and society.
2. An emphasis on style at the expense of substance and content.
3. The breakdown of the distinction between high culture (art) and popular culture.
4. Confusions over time and space.
5. The decline of the meta-narrative.

The key postmodern theorists discuss some of these concepts in detail. Connor (1989) cites the main theorists in the field as being Jean-François Lyotard, Frederic Jameson and Jean Baudrillard. Lyotard's work focuses on the function of narratives, rejecting the concept of universal meta-narratives, which he views as culturally imperialistic. Instead, he advocates cultural diversity and plural narratives. He also recognises the growing importance of the cultural industries and tourism to the promotion of cultural diversity: 'The world market does not constitute a universal history in the modern sense. Cultural differences are in fact encouraged even more, by virtue of the whole range of tourist and cultural industries' (1986: 62).

Baudrillard defines postmodernism as a world of 'simulations' and 'hyper-reality'. Western capitalist democracies have become economies based predominantly on the production of images and information. He defines the postmodern world in terms of codes, signs and symbols, which determine the nature and control of cultural and economic production. These are societies of the 'simulacrum' (identical copies without an original) where the simulated and the real are difficult to differentiate. The result is hyper-realism, a world where the unreal is presented as reality, and people are sometimes wont to confuse the world of culture and media with their real lives. Eco (1986) also describes hyper-reality as being pervasive, particularly in theme parks and museums. Baudrillard argues that tourists tend to thrive on 'pseudo-events' or inauthentic contrived attractions, and are isolated from the 'real' world (Urry, 1990).

Jameson argues that 'Historicism' has replaced history particularly within the heritage industry. Random elements of history are selected and pastiched or collaged and are then presented to the public as simulacra. Hewison (1991: 175) also describes how 'Heritage is gradually effacing history, by substituting an image of the past for reality'. Jameson also characterises postmodernity in socio-economic terms. His theories were predominantly Marxist, but, unlike earlier Marxist social theorists, he argues that cultural forms are no longer part of an ideological veil, which prevents real economic relations from being seen. Cultural production has become an inherently and openly commercial expression of pure capitalism. As stated by Urry (1990: 85) 'Commerce and culture are indissolubly intertwined in the postmodern'.

The complexity of these cultural debates cannot be overestimated, especially in a global and multicultural environment. As stated by Smith (1996: 43), 'There is no doubt that postmodernism has helped to incubate a serious analysis of the cultural dimensions of urban change that had hitherto been lacking', and Featherstone (1991: 11) states that 'Postmodernism is of interest to a wide range of artistic practices and social science and humanities disciplines because it directs our attention to changes taking place in contemporary culture.' Urry (1990: 82) describes how 'Postmodernism involves a dissolving of the boundaries, not only between high and low cultures, but also between different cultural forms, such as tourism, art, education, photography, television, music, sport, shopping and architecture.' The following section will discuss some of these relationships, focusing in particular on the experience of the 'post-tourist'.

## Postmodernism, tourism and the post-tourist

Urry (1990: 87) describes tourism as being the quintessential postmodern industry: 'Tourism is prefiguratively postmodern because of its particular combination of the visual, the aesthetic and the popular'. Rojek and Urry (1997: 3) write about the development of a 'post-modern cultural paradigm [which] involves the breaking down of conventional distinctions, such as high/low culture, art/life, culture/street life, home/abroad'. Postmodern tourism could be described as a form of 'pastiche tourism' (Hollinshead, 1997: 192), or 'collage tourism' (Rojek, 1997: 62).

The profile of the postmodern tourist is discussed by both Urry (1990) and Walsh (1992). They describe how many postmodern consumers receive much of their cultural capital through media representations, including travel. They cite Feifer (1985), describing the 'post-tourist' as one who does not necessarily have to leave the house in order to view the typical objects of the tourist gaze. The simulated tourist experience is brought into our living rooms through television travel shows, internet sites and software programs. As stated by Adair (1992: 24), 'Culture, in short, is something which "happens" to us increasingly at home.' Urry (2002: 83) refers to the concept of a 'three minute culture', which is characteristic of the media and its televisual influence. Bayles (1999: 166) uses (albeit cynically) the metaphor of the television to describe contemporary cultural consumption:

> It is now academic orthodoxy that all of culture – indeed, reality itself – is a torrent of images cut off from one another and from time, space, and meaningful reference and without emotional impact. To put the matter in nontheoretical parlance: Life is channel zapping.

McCabe (2002) argues that tourism has become such an established part of everyday life, culture and consumption that it is hard to differentiate it from other domestic and leisure activities. The tourist experience becomes little more than an exaggeration, enhancement or enrichment of everyday activities: 'Tourism represents a microcosm of everyday life, a magnifying glass through which the entire miscellany of life is distilled into a fragmentary week or fortnight' (McCabe, 2002: 70). However, Urry (2002) suggests that tourists are still essentially looking for difference when they travel. Craik (1997: 114) explores the relationship between home and abroad in more depth, suggesting that:

> Tourists revel in the otherness of destinations, people and activities because they offer the illusion or fantasy of otherness, of difference and counterpoint to the

everyday. At the same time the advantages, comforts and benefits of home are reinforced through the exposure to difference.

As the world becomes more globalised, the homogenisation and standardisation of cultural experiences and activities are perhaps inevitable; hence people may need to travel further afield in order to experience differences. Despite the appeal of virtual and simulated worlds, tourism keeps growing: 'tourism is growing at a time in which the effects of simulation have eroded our distinctions between elite and popular, reality and fiction' (Rojek, 1997: 70). The post-tourist can now benefit from the veritable smorgasbord of high and low cultural pursuits, indulging in both simultaneously. It is clear that our cultural preferences are determined by our own, personal sphere of interests and passions, and are often also linked to our individual, or sense of group identity. We are socially conditioned or educated to aspire to certain cultural activities rather than others, although in the end it is often simply a matter of individual or group preferences. For example, to some, the idea of a 'cultural' night out might consist of a visit to the theatre, opera or ballet, preceded by a formal dinner, intellectual discussion and a bottle of Chardonnay. To others, it might involve a couple of beers in the local pub, a few jokes or songs, followed by a football match or a pop concert. Some would argue that access to certain cultural activities is limited; for example, that the arts are too expensive for the masses to afford. However, a visit to the theatre or ballet can often cost the same or less than attending an average football match or pop concert.

Many of us will, of course, wish to pick and mix from the vast array of different cultural activities available to us. The simultaneous consumption of diverse cultural activities (e.g. high and low, traditional and contemporary, mass and elite) is becoming increasingly common, especially among the younger generations. For example, it is not unusual for bookshelves to be lined with the complete works of Shakespeare and Oscar Wilde on the one hand, and the latest Nick Hornby or Helen Fielding novels on the other. One side of a CD stand may hold Mozart or Beethoven symphonies, while the other displays the greatest hits of Madonna, George Michael or the Spice Girls. An interest in so-called high, elite or traditional cultural activities and low, popular or contemporary activities need not be mutually exclusive. In fact, the fusing of traditional and contemporary cultural forms is becoming a more common phenomenon in an increasingly globalised, postmodern world. For example, using the music of Pavarotti and Fauré respectively to introduce the 1994 and 1998 World Cup football championships has helped to bring classical music to new, often uninitiated audiences; classical violinists such as Nigel Kennedy or Vanessa Mae often play pop music as part of their repertoire; modern-day interpretations of Shakespeare plays have been produced with pop soundtracks for cinema audiences (e.g. Baz Luhrmann's *Romeo and Juliet*); *Poems on the Underground* in London has helped to bring classic literature to the masses as they take the train to work or to the pub; Stephen Daldry's 2000 film *Billy Elliott* challenged the prevailing 'lad culture' in Britain by tracing the fate of a boy ballet dancer from a traditional mining town in the north of England.

It can be seen that the changing and diversifying tastes of the modern-day consumer are being catered for *par excellence* by the travel market. Tourism is the quintessential global industry, fusing international travel with the desire for leisure and recreational activities of all kinds, and, increasingly, an interest in the multifarious cultures of the world. The average tourist today is likely to want to combine a visit to a beach with a weekend's shopping, a day or two of sightseeing, an evening at the theatre or a concert, followed by a couple of bars or nightclubs. Many tourists can no longer be as easily pigeon-holed into the 'mass tourist' (beach and clubbing) type, and the 'cultural' (sightseeing and arts event) type.

For the post-tourist, tourism has become a game: 'the post-tourist knows that they are a tourist and that tourism is a game or a series of games with multiple texts and no single, authentic tourist experience' (Urry, 1990: 100). Rojek (1993) sees the consumption experience as being accompanied by a sense of irony. He suggests that the quest for authenticity and self-realisation is at an end, and we are now in a stage of post-leisure and post-tourism. He describes the post-tourist as having three main characteristics. These are:

1 An awareness of the commodification of the tourist experience, which the post-tourist treats playfully.
2 The attraction to experience as an end in itself, rather than the pursuit of self-improvement through travel.
3 The acceptance that the representations of the tourist sight are as important as the sight itself.

## Authenticity and the post-tourist experience

The concept of authenticity is arguably a subjective attribution. Cohen (1988) argues that authenticity is a 'socially constructed concept' and that the meaning is negotiable. It is also a relative concept, as stated by Moore (2002: 55): 'One person's absolute fake is another's meaningful experience.' Getz (1994: 425) describes authenticity as:

> a difficult concept open to many interpretations, but [is] of great importance in the context of cultural tourism and particularly event tourism. Although some believe that authenticity is an absolute, determined by a complete absence of commoditization, many other theorists believe it is transitory, evolving and open to negotiation.

Jamal and Hill (2002) provide an excellent analysis of different typologies of authenticity. They differentiate between 'objective authenticity', which usually refers to traditional or historical sites or artifacts, and 'constructed authenticity', which may refer to staged events, moderated art objects, or artificially created cultural attractions. The category of 'personal authenticity' is perhaps the most complex and the least researched, but may refer to the emotional and psychological experience of travel, subjective responses to, and interpretation of sites and events experienced, or deeper existential aspects relating to personal meaning and identity. They conclude that:

> 'Authenticity' is neither a unified static construct nor an essential property of objects and events. It is better to approach it more holistically as a concept whose objective, constructed and/or experiential dimensions are in dialectical engagement with each other and with both the home and world of the tourist. Tourism becomes a metaphor for a changing, bio-political world in which (post)modernity, capitalism and globalization furnish complex meanings to authenticity and the authentic in everyday life.
>
> (Jamal and Hill, 2002: 103)

Turner and Ash (1975) describe how tourists are placed in a circumscribed world devoid of responsibility and protected from reality, which includes any sense of authenticity. Boorstin (1964) argues that tourists deliberately go in search of inauthentic experiences or 'pseudo-events', and that tourism has become responsible for rendering most events superficial or 'pseudo'. Of course, many tourists are escapist in their pursuit of leisure and entertainment, but many are also very well aware that the cultural experiences presented to them are far from being 'authentic'. There are often more important issues at stake for the average tourist than authenticity, such as entertainment and enjoyment. However, this is probably less likely to be the case for tourists who partake of indigenous cultural tourism. In many ways they go to great lengths to *avoid* the inauthentic. They conform to MacCannell's (1976) description of tourists as being like contemporary pilgrims who are in search of authentic experiences in other places and other times. In terms of the world around us, Goffman (1959) argued that the 'reality' of everyday living which we perceive is as staged as the cultural performances we attend. This is the concept of 'staged authenticity'. The real lives can only be found 'backstage', but as stated by Urry (1990) the gaze of the tourist will then become an obvious and unacceptable intrusion into people's lives which any sensitive tourist would want to avoid. Crick (1988) also contends that all cultures are staged and are therefore inauthentic to an extent. It has to be accepted that many tourists cannot be entirely sure whether or not a cultural performance is entirely authentic, whereas the performers should have a very clear idea. The problem arises when the tourist feels deceived or disappointed and the performer feels exploited, stereotyped or compromised. Clearly, local cultural performances and events should not be adapted or altered in such a way that they offend local sensibilities or compromise artistic integrity.

Getz (1994) refers to three perspectives of event authenticity: the anthropological perspective of the inherent cultural meanings of festivity and celebration; the planning perspective of community control and the mobilising of local resident support; and the visitor experience and perception. Community control is the main factor that determines how the event is presented and promoted. If local support and approval are achieved, then it is more likely that the tourist will be given an authentic and enjoyable experience.

However, the post-tourist is now apparently aware that the tourist experience is largely commodified, and that the quest for authenticity is somewhat futile. Craik (1997: 114)

**Plate 1.1** Tourists enjoy an abridged version of Kathakali in Kerala, India

Source: Author

describes this new tourist experience as 'an ego-centric pursuit, involving a fascination with self-indulgence and self-delusion through simulacra: approximations and analogues of "the real"'. Boniface and Fowler (1993: 7) state that:

> We want extra-authenticity, that which is better than reality. We want a souped-up, fantastic experience. We want simulation of life ways as we would wish them to be, or to have been in the past. As is clear, the travel industry knows it is dealing in dreams.

Myth and fantasy have always been central to the tourist experience. As stated by Rojek (1997: 52), 'Mention of the mythical is unavoidable in discussions of travel and tourism', and Tresidder (1999: 147), 'tourism at its most simplistic level is concerned with the production and consumption of dreams'. Kirschenblatt-Gimblett (1998: 144) describes how representation has become almost more important than the destination: 'The industry prefers the world as a picture of itself.'

Tourism packages now offer the tourist a whole range of facilities to accompany their visit to major sights and destinations. Tourism has become a much more integrated experience, no longer simply a focused quest for knowledge, self-improvement and authenticity of experience on a whistle-stop tour of 'must see' sights. Shopping, eating, drinking and evening entertainment are becoming as much a part of the tourism product as visiting the world's major monuments. Ironically, many of these activities can serve as a form of compensation for the disappointed tourist whose experience of the world's major sights fails to live up to the glossy media images and other forms of representation with which the tourist has been presented. Who, for example, would have imagined from the postcards and tourist brochures that the Mona Lisa would be so small, or that the Pyramids would be surrounded by urban sprawl, or that the Taj Mahal would be discoloured and viewed through a haze of smog?

Many sights are now really viewed as only second-hand images which bear little resemblance to reality. For example, Urry (2002: 55) describes how the Niagara Falls 'now stand for kitsch, sex and commercial spectacle. It is as though the Falls are no longer there as such and can only be seen through their images.' However, many tourists appear to be content to gaze upon what is familiar to them. Urry (2002) describes tourists as semioticians who are searching for signifiers that are familiar to them through other media (e.g. a typical English village, French chateau, German beer garden, American skyscraper).

Deception or disappointment appear to be an accepted part of the tourist experience (Rojek, 1997; Tresidder, 1999). De Botton (2002) suggests that travel rarely lives up to our expectations, partly because images are idealised and romanticised in art, literature and media, and reality is often a far cry from the idyll. Ironically, tourists are often quite taken aback if the landscape on which they gaze even so much as resembles a postcard image!

Second, we frequently have problems escaping from ourselves and the stresses and strains that we so desperately want to leave behind: 'The pleasure we derive from journeys is perhaps dependent more on the mindset with which we travel than on the destination we travel to' (de Botton, 2002: 246). Finally, many tourists are becoming weary of travel. Richards (2001a) suggests that people have effectively been overloaded with culture, and that many tourists are suffering from 'monument fatigue', wearied by the phenomenon of 'musterbation' (the apparent compulsion to visit 'must-see' sights!). The quest for new experiences, rather like the insatiable accumulation of material possessions in consumer societies, is a wearisome and ultimately unfulfilling process. De Botton (2002) notes that tourists often feel intimidated by guidebooks and fail to react spontaneously and subjectively to destinations and sights, but instead remain obedient to a prescribed route. Hence we fail to see the world clearly and to appreciate some of the smaller details of travel that could truly enhance our lives. Urry (2002) also notes that 'The contemporary tourist gaze is increasingly signposted. There are markers that identify the things and places worthy of our gaze.' Edensor (2001: 74) analyses this phenomenon in more depth, concluding that every potential space is effectively stage-managed and regulated: 'Thus there is a machinery of discursive, regulatory and practical norms which direct tourists' performances and often support their own understandings of how to behave.'

Despite this rather gloomy perspective on the contemporary tourist experience, it could be argued that many post-tourists are far from disappointed by their inauthentic experiences, and are drawn to 'hyper-real' attractions (e.g. theme parks, leisure centres or shopping malls), simulacra (e.g. Santa Claus' Lapland), or as described by AlSayyad (2001), sites of 'authentic fakery' such as Las Vegas or manufactured heritage theme parks of 'fake authenticity'. The following section will discuss the nature of such attractions in more detail.

## The development of popular cultural tourism attractions

Ritzer and Liska (1997) describe how McDonald's and Disney have become as much symbols of the postmodern tourist landscape as any other cultural icons. They are also symbols of increasing globalisation. As stated by Warren (1999: 123) in her discussion of Disney theme parks: 'Disneyland Paris speaks to a fear lurking deep in social theory: that the nation-state is obsolete, about to be eclipsed by the multinational corporation.' Ritzer and Liska (1997) suggest that their homogenising omnipresence – the 'McDonaldisation' and McDisneyisation' of the world – has undermined somewhat the fundamental reason for tourism, which is to experience something new and different. Kirschenblatt-Gimblett (1998: 9) describes how the world has become a kind of museum of itself: 'Tourists travel to actual destinations to experience virtual places.' Theme parks, shopping malls, fast food have all become part and parcel of the same postmodern consumption experience. They cite Barber (1995: 97) who summarises the concept of 'McWorld' as:

> an entertainment shopping experience that brings together malls, multiplex movie theatres, theme parks, spectator sports arenas, fast-food chains (with their endless movie tie-ins), and television (with its burgeoning shopping networks) into a single vast enterprise that, on the way to maximising its profits, transforms human beings.

These are the ultimate integrated, 'inauthentic' experiences, and this would all suggest that there is increasingly very little differentiation between the leisure, recreation and tourism experiences of our lives (see Urry, 1990).

Rojek (1993) identifies four kinds of tourism and leisure attractions which feature in the landscape of postmodernism. These are:

1 Blackspots;
2 Heritage sites;
3 Literary landscapes;
4 Theme parks.

Blackspots refer to the commercial development of sites of atrocity, such as graves, war zones, massacre, assassination or accident sites. Many of these have featured in movies (e.g. *Schindlers List, The Bridge over the River Kwai*); therefore visitors are often motivated to visit such sites as much because of their association with a movie as for their historical significance. It is worth considering how far the post-tourist is able to differentiate between fiction and reality when history and media are merged. As discussed earlier, in Baudrillard's world of simulation and hyper-reality, the boundaries are sometimes blurred, and people find it difficult to distinguish between fantasy and reality. Urry (1999: 34) suggests that 'Tourism provides the opportunity to live outside both time and space for a limited period.'

There is currently an increasing trend towards 'movie induced tourism' (Riley *et al.*, 1998) whereby tourists visit fictional landscapes which were used as settings for films. Similarly, many heritage sites are created in order to re-enact or to stage the past. In some cases, history is partially invented and then presented as heritage. Rojek (1993) gives the example of Robin Hood country which is marketed heavily in Nottinghamshire, but records suggest that Robin Hood existed only in folklore rather than reality. This is a form of mythical tourism. The same is true of some fictional literary landscapes (e.g. Catherine Cookson country or Thomas Hardy's Wessex) which have been developed as tourist attractions (Box 1.1). There is a kind of symbiosis between the writer, their novels, the place they lived and the settings for their novels. As stated by Herbert (1995b: 33), 'Literary places are the fusion of the real worlds in which the writers lived with the worlds portrayed in the novels. Any distinction is unlikely to be made in the minds of visitors.'

Although the post-tourist is perhaps less concerned about authenticity than his or her predecessor, the commodification of history is still a cause for concern. For example, a number of authors have criticised the way in which many heritage sites and landscapes have become more and more like theme parks (e.g. Hewison, 1987; Walsh, 1992; Kirschenblatt-Gimblett, 1998). This is due largely to the postmodern pastiche of reality and fiction, and history and media, which can result in a kind of simulacra or 'time capsule'. However, Walsh (1992: 103) does differentiate between types of attraction, distinguishing between those that offer pure entertainment and those that offer a combination of education and entertainment, or the 'edu-tainment' phenomenon (Urry, 1990). Commercial theme parks are perhaps the ultimate postmodern tourist attraction, and they arguably differ greatly from themed heritage attractions:

> Such attractions vary enormously in their emphasis on the education/entertainment ratio. There are the Disneylands, the Alton Towers, the Camelots and the American Adventures, which like postmodern architecture, are concerned with imagineering [*sic*] projects in their most basic form – the development of an environment constructed through historical surfaces, a context of superficial spectacle-consumption and entertainment.
>
> (Walsh, 1992: 103)

Craik (1997) suggests that artificial theme parks are much more appealing than many themed heritage attractions or museums because they can offer the visitor a more exciting, entertaining and integrated experience.

## Theme parks: the ultimate postmodern experience?

Philips (1999: 93) describes how 'the theme park is a space unequivocally devoted to pleasure'. This differentiates it from heritage sites which generally purport to being educational in some way. This is not to say that visitors will learn nothing of value in the theme park, but this is not its primary aim.

Craik (1997: 115) describes theme parks as the ultimate 'tourist bubble' (a safe, controlled environment) out of which they can selectively step to 'sample predictable forms of experiences'. Theme parks, especially Disney parks, are viewed as safe, secure and

## Box 1.1 Literary, media and film tourism in the UK

Literary tourism is big business in the UK. Stratford-upon-Avon as the birthplace of Shakespeare is perhaps the best example of this. Dublin is commonly associated with Oscar Wilde and James Joyce, Canterbury with Chaucer, the Lake District with William Wordsworth, and Dorset with Thomas Hardy, to name but a few. Pocock (1987, 1992) provided an analysis of visitors to Brontë country in Yorkshire and Catherine Cookson Country in South Tyneside. In both cases visitors are encouraged to visit the worlds of the novels and those of the writers, but he suggests that many visitors are unlikely to be able to distinguish one world from the other.

However, this is not to say that visitors to literary heritage sites are ill-informed. In terms of motivation, Pocock (1992) described many visitors as 'literary pilgrims' because of their desire to be educated about the life and works of the author. Herbert (1995b) focuses on visitors to Jane Austen's residence at Chawton, Hampshire, concluding that most visitors were motivated by a genuine literary interest. Squire (1993) describes how visitors to Beatrix Potter's home in the Lake District seem to be attracted partly by the nostalgia associated with a kind of rural idyll, as well as by the author's works. Craik (1997: 116) suggests that the diversity of the Potter industry had somehow detracted from the main literary theme: 'the Potter attractions – both authentic and constructed – were simply the backdrop or catalyst for the pleasures, connections and projections that individual tourists had derived from their visit.' Literary tourism can thus be used as a magnet or a catalyst for the development of rural or urban tourism.

Many other areas are also being promoted and commoditised increasingly because of their association with famous writers. For example, Rochester has a themed Dickens Trail, and hosts a Dickens Festival. Like the Catherine Cookson Trail in South Shields, many of the visited landmarks are fictional. There are a number of restaurants, shops and bars which are themed according to the Sherlock Holmes myth around Baker Street in London. He was, of course, a fictional detective, but as stated by Rojek (1993: 153), 'Thus is fiction co-opted in the service of commerce and myth mingles with reality'. The same is true of Thomas Hardy's 'Wessex' and the fictional town of 'Casterbridge', whereby several streets in the town of Dorchester (which was Hardy's model for the fictional town) have been named after Hardy's characters (ibid.).

The proliferation of costume dramas has helped to bring the works of such authors to life, and visitors are becoming more interested in visiting film locations as well as the landscapes associated with the authors and their works. For example, the British Tourist Authority has produced a *Movie Map* which includes numerous film locations, such as *Pride and Prejudice*, *Braveheart*, *Remains of the Day*, *Four Weddings and a Funeral*, and *Brassed Off*. 'Movie-induced tourists' (Riley *et al.*, 1998) are drawn to film locations largely because of the uniqueness of the landscape. This can create problems if the destination and its people are not adequately prepared for an influx of tourists, and commercialisation and commodification of place can become a major issue for locals. Problems may also arise if a storyline places an event in a named destination, yet filming takes place elsewhere. Authenticity can then become an issue for promoters and interpreters; however Busby and Klug (2001) suggest that many visitors may not mind whether or not the area visited is a genuine location. Once again, the boundaries between fiction and reality can become nebulous.

dependable: 'Disney World is highly predictable. . . . Indeed, Disney theme parks work hard to be sure that the visitor experiences no surprises at all' (Ritzer and Liska, 1997: 97). Zukin (1995: 55) describes theme parks as follows:

> While it is relished as a collective fantasy of escape and entertainment, the theme park is a tightly structured discourse about society. It represents a fictive narrative of social identity – not real history, but a collective image of what modern people are and should be – and it exercises the spatial controls that reinforce this identity.

The theme park becomes almost a microcosm of the world, or at least an idealised version of it.

Rojek (1993) describes how there are more themes than theme parks, but he describes all theme parks as being based around the 'meta-themes' of velocity and time–space compression. By this he is referring to the 'thrill factor' provided by fast rides, and the way in which time and space are dissolved in a diversity of experiences and spectacles. Visitors may have the experience of moving through time or travelling across continents, all within the space of a few hours. Philips (1999) describes a theme park as a space without clocks, as well as being a bounded space which is located outside the familiar environs of everyday life:

> The theme park explicitly offers a 'phantasmagoria': it celebrates the fact that it can bring together 'absent others', and revels in the exoticism of its attractions. The theme park is a space which is unapologetically penetrated by influences quite distant from their geographical location, and which distances itself from the actual locale.
>
> (Philips, 1999: 106)

In this sense, the theme park is a little like the simulated worlds in cyberspace. Philips (1999) describes how theme parks tend to be constructed around a number of specific themes which often correspond to popular literary genres, such as science fiction, fairy or folk-tales, and explorers and treasure islands. Again, this blend of fiction and fantasy is what gives the theme park its main appeal. The Disney concept arguably epitomises all that a successful theme park should be. Warren (1999: 109) describes Disney theme parks as 'the tourist meccas of the late twentieth century', and Zukin (1995: 50) states that 'Disneyland and Disney World are the most important tourist sites of the late 20th century'. They are visited by more people than many of the major monuments of the world. Although Disney is most commonly depicted as epitomising American culture, the brand is truly global, and the Disney Company is one of the most successful multi-media corporations of all time. The name is synonymous with global enterprise and initiative. Although there was initial concern that such a phenomenon could not be transported easily from its American home, Disneyland Paris has proved to be an unprecedented success in Europe. For example, more people now visit Disneyland Paris every year than the Eiffel Tower!

Zukin (1995) suggests that Disney World represents a privatised, sanitised, aestheticised, idealised world in which people can take refuge from the harsh realities of the outside world. However, she also suggests that there is much to be admired about the way in which Disney has managed effectively to create an environment that is in some ways more real than hyper-real:

> Like all the world's fairs that preceded it, this is a visual narrative for a compact tourism of exotic places. And it is a world's fair brought to you by a world-class corporation, whose references to its own cultural products are so entangled with

> references to those of real places that Disney World is indistinguishable from the real world.
>
> (Zukin, 1995: 58)

Disneyland is a tourist bubble which offers people a safe and predictable experience within a secure environment. For many visitors, especially children, it is the ultimate leisure experience. It is a world of fantasy and escapism, combining dreamscapes with simulations of real places, a curious blend of fiction and reality. However, Baudrillard suggests cynically that Disneyland presents itself as an imaginary space so as to conceal the fact that it reflects directly the characteristics of the real country in which it is located.

Similarly, Zukin (1995: 188) describes how late twentieth-century shopping malls have become viewed by many social theorists as being 'primary public spaces of postmodernity'. The public culture of mass consumption has found its home in the multiplex shopping mall which aims to combine retail therapy with other leisure and recreational pursuits, such as cinema, roller-skating or ten-pin bowling.

It is clear that the shopping centre, like the theme park, is becoming a symbol of global consumerism, embodying Barber's (1995) concept of 'McWorld'. Under one roof, visitors have the chance to experience a wide range of global fashion and international cuisine. Sarup (1996) refers to the shopping mall as an alternative life-world in which one can create alternative selves or identities:

> Through the market, one can put together elements of the complete 'Identikit' of a DIY self. These merchandised identities come complete with the label of social approval already stuck on in advance.

Urry (2002: 135) describes how:

> Malls attract their share of 'post-shoppers', people who play at being consumers in complex, self-conscious mockery. Users should not be seen simply as victims of consumerism, as 'credit card junkies', but also as being able to assert their independence from the mall developers. This is achieved by a kind of tourist *flanerie*, by continuing to stroll, to gaze and to be gazed upon.

The combination of shopping with other recreational pursuits suggests that the boundaries between retail and leisure are becoming more blurred, as are the boundaries between leisure and tourism. For example, many leisure complexes are now being developed as self-contained holiday destinations. The Center Parcs concept is a good example of this, and it is interesting to note that many of these complexes have been built in the colder climates of northern Europe (e.g. the Netherlands, Belgium, Britain and France). Once again, it is a form of simulated attraction – the creation of a kind of displaced tropical paradise in which those deprived of sunshine and beaches can indulge their fantasies. The world's geography can be experienced vicariously as simulacrum through themed environments such as Expos, which contain national displays of cultural activities (Urry, 2002).

It is worth questioning, of course, how far such attractions may be described as cultural tourism attractions. Clearly, if culture is defined broadly enough, then all aspects of leisure and entertainment could be said to constitute components of the product. However, it is suggested in this book that the majority of cultural tourists are keen to escape from the so-called tourist bubble and to engage in authentic experiences wherever possible. Hence they perhaps conform less closely to the concept of the post-tourist. Chapter 2 explores some of these issues in more depth.

## Conclusion

This chapter has considered the impacts of globalisation on culture and travel patterns, particularly in a Western, urban context. It is ironic that the standardisation of culture in the 'global village' is sometimes probably the very factor driving its citizens to travel more than ever! This is the only way in which they can experience cultural differences and authentic 'reality'. In the globalised environment of high technology, media and communications, virtual reality, hyper-reality and simulations, it is difficult to differentiate between fiction and fact. Perhaps, in travelling to a place where life is lived at a slower pace and some remnants of tradition can be found, the global citizen finds his or her own respite from the frenetic intensity of global living. Tourism has been viewed traditionally as an escapist phenomenon, one which is based predominantly on elements of fantasy, fictions and myths. Therefore, who would have thought that we would reach a stage where our everyday lives would become even more 'unreal' than the tourist destinations we visit? This certainly gives the post-tourist something to think about.

# 2 Reconceptualising cultural tourism

> If all tourism is culture and culture has become tourism where does that leave the study of cultural tourism?
>
> (Richards, 2001b: 2)

## Introduction

The aim of this chapter is to provide an overview of the development of cultural tourism, including an analysis of some of the factors that have shaped its development. The chapter will discuss the development of the cultural tourism product and the diverse range of cultural activities that are now considered to be part of this growth phenomenon. It will be argued that cultural tourism can no longer be considered as a special interest or niche sector, but instead as an umbrella term for a range of tourism typologies and diverse activities which have a cultural focus. The motivation for such travel will also be discussed, focusing on the changing nature of patterns of consumption and diversifying typologies of the cultural tourist as the phenomenon apparently shifts from 'niche' to 'mass'.

## The development of cultural tourism

In the light of the discussions in Chapter 1, it is evidently difficult to establish a universally valid definition of cultural tourism. It is relatively easy to fall into the trap of using terms such as 'heritage tourism', 'arts tourism', 'ethnic tourism' or 'indigenous tourism' almost interchangeably. It will be argued in this book that cultural tourism is indeed broad in its remit, but that there is perhaps a need for differentiation within the cultural tourism sector. Hence, such forms of tourism would all become subsets of cultural tourism or niche components within a diverse sector.

Richards (2001a: 7) suggests that cultural tourism covers:

> not just the consumption of the cultural products of the past, but also of contemporary culture or the 'way of life' of a people or region. Cultural tourism can therefore be seen as covering both 'heritage tourism' (related to artefacts of the past) and 'arts tourism' (related to contemporary cultural production).

Clearly, the notion of past and present implies that cultural tourism is based on both the history and heritage of a place and its people, as well as on their contemporary lives. Zeppel and Hall (1992) divided cultural tourism into the subsets of heritage and arts tourism. Arts tourism might be considered to be a more contemporary phenomenon, being located predominantly in the present, and it is arguably more experiential than heritage tourism.

However, in many ways, the arts and heritage are inextricably linked, and it is almost impossible to distinguish between them, particularly in the context of indigenous communities where the distinction between past, present and future is not as clear-cut or linear as in Western societies. Many traditions within the arts form a distinctive component of the heritage of a people or a place. This is especially true of crafts production or festivals. Even in historic cities (for example, in Italy), it is difficult to distinguish between the heritage and arts component of the cultural tourism product. Historic buildings host art exhibitions, theatre and opera take place in ancient amphitheatres, festivals and events are based in heritage streets. Boundaries are nebulous, and distinctions are not always possible or indeed useful.

Cultural tourism is as much based on experiencing as it is on seeing; hence Williams' definitions of culture as a whole way of life as well as the arts and learning are particularly relevant here. Past definitions of cultural tourism have perhaps placed too much emphasis on cultural tourism as a form of arts or heritage tourism in its narrowest sense; for example, visiting museums, monuments, galleries and theatres. In 1991, the European Association for Tourism and Leisure Education and Research (ATLAS) launched a Cultural Tourism Research Project for which they defined cultural tourism as:

> Technical Definition: All movements of persons to specific cultural attractions, such as museums, heritage sites, artistic performances and festivals outside their normal place of residence.
>
> (Richards, 1996: 24)

Of course, much of ATLAS's original research took place in a European context; hence the emphasis does tend to be placed on a cultural tourism product that favours arts and heritage tourism above indigenous or ethnic tourism. However, their conceptual definition takes us closer to the idea of culture as a way of life:

> Conceptual Definition: The movement of persons to cultural manifestations away from their normal place of residence, with the intention to gather new information and experiences to satisfy their cultural needs.
>
> (Ibid.)

Although this is a rather broad definition, it does imply that cultural tourists are interested in the more experiential aspects of culture. In an international context, particularly in the context of indigenous or ethnic tourism, the way of life of a people is a central focus. However, once again, it is difficult to distinguish between these rather nebulous concepts. Both forms of tourism assume that the traveller is motivated primarily by first-hand, authentic or intimate contact with people whose ethnic or cultural background is different from their own. One way of making a distinction might be to argue that indigenous tourism implies visiting native people in their own habitat which is different from that of the tourist, whereas ethnic tourism could refer to engaging in the cultural activities of a minority group within the tourists' own society. For example, for the purposes of this book, a distinction is made between indigenous cultural tourism and ethnic cultural tourism. The former refers to the lifestyles and traditions of tribal groups living within fragile and remote environments, often in post-colonial developing countries, whereas the latter refers to the arts and culture of ethnic minority groups, immigrants and diasporas living largely within post-imperial Western societies.

Eagleton's (2000) concerns that definitions of culture are both too broad and too narrow are perhaps borne out in the field of postmodern cultural tourism studies. It is here that distinctions between high and low culture are being broken down, and emphasis is being placed increasingly on popular or mass culture. This is equally true of the heritage and museum sectors where representation is becoming a key issue, and the histories of previously marginalised groups are being recognised. The rejection of so-called 'grand narratives' has meant that the discourses of the working-classes, women, and minority or ethnic groups are now being heard. Historicity is becoming a more valid concept than aesthetics rendering the social history and industrial heritage of the working classes as important as political history or the bourgeois heritage of royalty, for example. Inclusion, access and democracy are the new buzzwords, and the underlying concepts are important in defining the shape of the future, not just in terms of cultural development, but in terms of all the concomitant political and social struggles which surround it.

The impact that such developments have had on redefining cultural tourism are quite significant. The following list suggests a comprehensive typology of cultural tourism, but it is recognised that the concept of culture as almost everything that we are and everything we do is problematic unless we differentiate between the activities that are contained within it.

- Heritage sites (e.g. archaeological sites, whole towns, monuments, museums).
- Performing arts venues (e.g. theatres, concert halls, cultural centres).
- Visual arts (e.g. galleries, sculpture parks, photography museums, architecture).
- Festivals and special events (e.g. music festivals, sporting events, carnivals).
- Religious sites (e.g. cathedrals, temples, pilgrimage destinations, spiritual retreats).
- Rural environments (e.g. villages, farms, national parks, ecomuseums).
- Indigenous communities and traditions (e.g. tribal people, ethnic groups, minority cultures).
- Arts and crafts (e.g. textiles, pottery, painting, sculpture).
- Language (e.g. learning or practice).
- Gastronomy (e.g. wine tasting, food sampling, cookery courses).
- Industry and commerce (e.g. factory visits, mines, breweries and distilleries, canal trips).
- Modern popular culture (e.g. pop music, shopping, fashion, media, design, technology).
- Special interest activities (e.g. painting, photography, weaving).

The profiles of tourists who engage in these different forms of cultural tourism are likely to be quite different, in many ways. It is therefore interesting to consider how far the differentiation of the cultural tourism product is a significant consideration in commercial terms.

## The cultural tourism product

During the 1990s, cultural tourism was identified as one of the major future growth areas in Europe (Zeppel and Hall, 1992). The WTO (1993) estimated that 37 per cent of all international trips would have a cultural element, and that this figure would increase annually by 15 per cent to the end of the century. Of course, in its broadest sense, cultural tourism cannot be considered to be a niche form of tourism at all. As global tourism continues to grow, it might be assumed that cultural tourism is growing in equal proportion. Indeed, Richards (2001b) suggests that it is inevitable that cultural tourism appears to be growing, because more and more tourist attractions are now being defined as 'cultural', and, as tourism grows internationally, there are bound to be more visits to cultural attractions.

In terms of tourist profiles, Bywater (1993) distinguished between 'culturally motivated', 'culturally inspired' and 'culturally attracted' tourists. 'Culturally motivated' tourists represent a small but commercially desirable market segment, since they tend to be attracted to a destination chiefly for cultural reasons. They are generally high-income visitors who spend several nights at a destination (e.g. visitors to the Salzburg Music Festival). 'Culturally inspired' tourists are attracted to internationally renowned cultural and heritage sites (such as Venice, Versailles, or the Alhambra in Granada). Although some of this group will be 'culturally motivated', many tend to spend short periods of time visiting major cultural destinations, and are not motivated easily to return to the same destination twice, to stay in one place for longer or to visit minor destinations instead. Unfortunately, for these reasons, this group tends to be chiefly responsible for many adverse impacts, especially in terms of the environment, where carrying capacity is exceeded at major heritage sites. 'Culturally attracted' tourists can also pose a similar threat, in the sense that they represent a major day-trip market, visiting cultural attractions or attending cultural events because they happen to be in the area. However, the distinction between these three segments is by no means clear-cut, and at other times of the year, 'culturally attracted' tourists may fall into one of the other two categories of cultural tourist.

The BTA (2002: 4) confirms this development, stating that:

> The distinction between people who consider themselves 'cultural tourists' and those who don't specify a particular interest in the arts or cultural tourism, are blurring. The majority of tourists enjoy some element of cultural tourism during their visit, which could range from going to an exhibition in an art gallery or museum, following a literary or film trail to enjoying a musical or theatrical performance.

The BTA estimate that around two-thirds of tourists in the UK who do not take a culturally specific holiday still visit a museum during their holiday. This suggests that cultural tourism is more mainstream than previously imagined. Richards (2001b) estimated that three-quarters of tourists in Europe visit a cultural attraction even if they do not consider themselves to be on a cultural holiday. Tour operators have consequently responded to this by developing packages which combine a number of activities, some of them cultural, but others based purely on entertainment, fun, or relaxation. For example, Kuoni offer long-haul packages to Thailand, which combine beach stays on islands with city shopping and nightlife in Bangkok, and hilltribe trekking in Chiang Mai. The latter activity is particularly popular with cultural tourists seeking authentic contact with local people, as it generally involves homestays or visits.

ATLAS cultural tourism research has shown that over 50 per cent of cultural tourists in Europe tend to have some form of higher education compared with 21 per cent of the EU population as a whole. They also tend to have professional occupations and earn considerably higher salaries than the average European (Richards, 2001b). However, as definitions of cultural tourism become more and more inclusive, these profiles may change over time. Nevertheless, care must still be taken to differentiate between activities which are deemed cultural as opposed to recreational or leisure-based, especially for the purposes of research and data collection. As stated earlier, there is perhaps scope for more differentiation between typologies of cultural tourism.

Richards (2001c) notes the development of the creative industries and the parallel development of a form of 'creative tourism'. He defines 'creative tourism' as tourism involving active participation by tourists in the creative process. This could include some of the more active forms of special interest tourism such as cookery, painting, photography, or arts and crafts holidays. Many people who do not have time for creative pursuits in their everyday lives are increasingly undertaking such activities on holiday. He suggests that creative tourism is being pursued particularly by those destinations which cannot compete on the basis of their cultural and heritage resources (for example, de-industrialising cities or rural areas).

There has clearly been a growing emphasis on the importance of creative industries to the global economy. In a world of high technology, media and communications, it is inevitable that modes of cultural production are changing. The cultural industries (e.g. the arts, heritage and museums) are succumbing to the usage of new and interactive technologies and media to enhance their products. Even artists themselves are now increasingly using photography, video footage and new media to express their ideas. Within this context, it could certainly be argued that the boundaries between art and media, culture and commerce, and fiction and reality are being gradually eroded.

## Demand and motivation for cultural tourism

In his insightful analysis of the motivations that underpin our desire to travel, de Botton (2002) suggests that like Baudelaire, we may be forever clamouring to be where we are not, seeking escapism or 'getting away from it all'. The destination may be largely irrelevant, hence a form of travel that may be more typical of 'psychocentric' package tourists rather than cultural tourists who tend to be of a more 'allocentric' disposition (Plog, 1974). Alternatively, we may be drawn to exoticism like Flaubert who was obsessed with travel to the Orient:

> In the more fugitive, trivial association of the word exotic, the charm of a foreign place arises from the simple idea of novelty and change . . . we may value foreign elements not only because they are new, but because they seem to accord more faithfully with our identity and commitments than anything our homeland could provide.
>
> (de Botton, 2002: 78)

Sarup's (1996) suggestion that travel allows us to enjoy and exploit simultaneously the exotic difference of 'the Other' while discovering our own identity is perhaps borne out here. Travel can help to bring us into contact with our true selves. Wang (2000) suggests that many tourists are more likely to be in search of their own 'existentially' authentic selves rather than seeking 'objective' authenticity. The boredom, lassitude or monotony of everyday life that may hinder our ability to feel authentic in an existential sense are often temporarily removed. This craving for difference and exoticism is perhaps stronger in the case of cultural tourists who will actively seek out remote locations, unusual experiences or close and authentic contact with indigenous groups. Seaton (2002: 162) suggests that the notion of the tourist gaze (Urry, 1990, 2002) should not be only about tourists seeing the world around them, but also about perceptions of self: 'Tourism is at least as much a quest *to be* as a quest to see.' Of course, the notion of escapism through travel could apply as much to escapism from self as escapism from place or routine, but, as noted by de Botton (2002), one of the barriers to the enjoyment of travel is the fact that we cannot easily get away from ourselves and our persistent worries. Similarly, Edensor (2001: 61) states that:

> Rather than transcending the mundane, most forms of tourism are fashioned by culturally coded escape attempts. Moreover, although suffused with notions of escape from normativity, tourists carry quotidian habits and responses with them; they are part of their baggage.

However, many tourists still crave the enhancement rather than the avoidance of self, particularly cultural tourists. They subscribe to Nietzsche's view of travel that it should be a constant process of knowledge-seeking and self-improvement. This would, of course, conform to the rationale behind the Grand Tour of the seventeenth and eighteenth centuries which was predominantly an educational and cultural experience. Many modern-day cultural package tours appear to emulate this philosophy. Seaton (2002) describes the process of 'metempsychosis' whereby tourists engage in repetitive or ritualistic behaviour, often following in the footsteps of famous figures on their travels. Certainly, many forms of cultural tourism, such as literary, media or film tourism, could be described as metempsychotic, as could certain forms of heritage tourism; for example, those that include re-enactments, or tours with a mythical element to them. He also discusses the concept of 'metensomatosis', or the process of temporary role-playing whereby tourists adopt multiple personae: 'The tourist is . . . typically a multipersonae traveller, a polyphrenic *bricoleur* whose tourism enactments are based on representations of what others have been in the past' (Seaton, 2002: 159). He argues that the act of role-playing with social peers in a new place is, in some cases, more significant and more common than interaction and engagement with local people. Hence, we return to the concept of self and existential authenticity being just as important as objective authenticity and gazing on 'the Other'. It is only by analysing the interaction between self and the world that we can truly understand tourist motivation. As stated by Graburn (2002: 31):

> Tourism, like life itself, can be represented as a journey. Indeed, the relationship between an inner and an outer metaphor may be the key to understanding tourists' motivation, expectations and satisfactions.

Although de Botton (2002) suggests that most destinations fail to live up to our somewhat idealised or romanticised expectations, or are at least different from what we expected, the lure of travel is not always dampened by such disappointments. The human instinct for novelty and difference is fuelled by the travel industry: its glossy brochures, and promises of an idyllic present, which rarely fail to capture our imaginations. Urry (2002: 13) describes the insatiability that appears to underpin our quest for travel:

> Peoples' basic motivation for consumption is not . . . simply materialistic. It is rather that they seek to experience 'in reality' the pleasurable dramas they have already experienced in their imagination. However, since 'reality' rarely provides the perfected pleasures encountered in daydreams, each purchase leads to disillusionment and to the longing for ever-new products. There is a dialectic of novelty and insatiability at the heart of contemporary consumerism.

Similarly, Sarup (1996: 128) states that 'Under postmodern conditions, there is the exhilarating experience of ever new needs rather than the satisfaction of the still-existing ones.'

In many ways, the cultural tourism industry with its constantly diversifying products helps to feed this insatiability. Not only are there now generic categories of cultural tourism, but there are also various subsets of these categories, as the following section will attempt to demonstrate through an analysis of cultural tourist typologies.

## Typologies of the cultural tourist

In his analysis of the tourist as a metaphor of the social world, Dann (2002: 82) implies that the profile of the postmodern tourist is likely to be closer to that of a pyschocentric hedonist than an allocentric 'traveller':

> Just as modernity had its metaphor of 'the traveller', seeking the rational goal of educational improvement, the moral path of spiritual renewal, the scientific and imperialistic exploration of unknown territories, so too did postmodernity seize upon the tourist as connotative of a dilettante life of fun in the sun and hedonism *ad libitum* in placeless destinations where the 'other' was cheerfully ignored in favour of the unbridled pursuit of individualism *sans frontières*.

Chapter 1 discussed some of the complexities of defining 'post-tourism' and the profile of the 'post-tourist', and questioned whether the majority of cultural tourists could be said to conform to this profile. Of course, this would depend largely on how broadly cultural tourism is defined, but this chapter has suggested that many cultural tourists are orientated towards arts and heritage, and that some leisure and recreational activities would fall outside the definitions adopted. The following list provides a brief comparison of the perceived profiles of both the post-tourist and the cultural tourist:

| **The post-tourist** | **The cultural tourist** |
|---|---|
| • Enjoys simulated experiences, often in the home. | • Keen on personal displacement and the notion of 'travelling'. |
| • Little differentiation between tourism, leisure and lifestyle. | • Actively seeking difference. |
| • Acceptance that there is no true authentic experience. | • Seeking objective authenticity in cultural experiences. |
| • Treats the commodification of the tourist experience playfully. | • Concerned with existential authenticity and enhancement of self. |
| • Ironic detachment from experiences and situations. | • Earnest interaction with destinations and inhabitants. |
| • Little interest in differentiating between reality and fantasy. | • May have idealised expectations of places and people. |
| • Interested in 'hyper-real' experiences. | • Interested in 'real' experiences. |
| • Acceptance of representations and simulacra. | • Disdain for representations and simulacra. |

Although it can be difficult to generalise about the profile and motivations of the average cultural tourist, this list suggests that there are significant differences between the interests, expectations and motivations of cultural tourists compared to post-tourists. The majority of cultural tourists will be keen to experience new and different places, and part of the pleasure of their experience will be derived from the process of travelling itself. Journeys are often not seen as a means to an end as they are in the case of package tourists, but as an exciting form of personal displacement, which affords new sights, sounds and smells. The use of local transport will be frequent, and many cultural tourists (especially backpackers) will often take great delight in being sandwiched between locals and their sacks of rice or grain, or their entourage of goats or chickens. Most cultural tourists are likely to be on some kind of quest for authenticity, either in terms of self-improvement or in terms of the sites, communities and activities that they engage with or in. They may see themselves as some kind of contemporary pilgrim on a spiritual quest, or as a 'metempsychotic' figure retracing intellectual or exploratory paths. They will want to engage fully with the destinations they visit and to interact with local inhabitants. Tourism will not be taken lightly or treated with ironic detachment and they will not want to witness forms of exploitation, commodification or 'fake authenticity'. They will more than likely be irritated by false representations of destinations and their people, and they will be less than keen to visit sites that are simulacra (e.g. heritage theme parks, model villages, staged events).

However, as discussed in Chapter 1, it is no longer appropriate to pigeon-hole tourists into one category or typology. As stated by Edensor (2001: 59), 'Typologies can identify regularities, but should be conceived as describing different tourist practice rather than types of people, as roles adopted rather than social categories made manifest.' Expanding on his concept of role-playing, Seaton (2002) suggests that transient rather than fixed typologies should be used as a way of perceiving tourist motivation. He suggests a number of personae that might be adopted or emulated by the tourist. Those that are most appropriate to the cultural tourist might include:

- *The dilettante/aesthete*: origins with the European Grand Tourist visiting museums, galleries, and other cultural sites and landscapes.
- *The antiquarian heritage seeker*: particularly interested in the classical past, history and archaeology.
- *The explorer-adventurer*: influenced by the development of mountaineering, hiking, trekking and backpacking.
- *The religious pilgrim and spiritual seeker*: visitors to pilgrimage destinations, and in particular Western tourists seeking spiritual enlightenment through Eastern religions.
- *The festival charivariist*: attraction to festivals or 'fests' which allow a certain degree of escapism and lack of constraint.
- *The litterateur*: interested in the homes and landscapes of literary figures.
- *The epicurean*: a bon viveur, gastronome or wine buff.
- *The natural and social scientist*: an interest in rural heritage or an ecotourist.

(Adapted from Seaton, 2002)

It is indeed interesting to characterise tourists according to the roles they play and the personae they adopt, although it could be argued that most tourists do not consciously role-play. According to more traditional typologies of the tourist, cultural tourists might fall variously into the categories of 'explorer' or 'drifter' (Cohen, 1972) or 'elite', 'off-beat' or 'unusual' (Smith, 1989). Although 'unusual' tourists tend to prefer the tourist bubble to striking out alone, they nevertheless adapt reasonably well to local culture. Sharpley (1994) refers to five categories of tourist experience: recreational, diversionary, experiential, experimental and existential. The first two categories of visitors are mainly seeking relaxation and escapism, whereas the last three categories will wish to immerse themselves either partly or fully in local culture and society, and authentic experiences will be sought.

Although it is ultimately misleading to suggest that generic typologies of cultural tourist exist, it is perhaps interesting to consider the different activities in which cultural tourists engage (clearly there would be inevitable overlaps between some of these categories) (see Table 2.1).

## Issues in cultural tourism studies

The following section outlines some of the generic categories of cultural tourism that are discussed in this book. It is recognised that many of these categories could be divided into further subsets (as demonstrated in the previous typologies); hence each chapter attempts to recognise issues relating to these subsets, even if there is insufficient scope to discuss them all in depth.

**Table 2.1** ***A typology of cultural tourists***

| *Type of cultural tourist* | *Typical places/activities of interest* |
|---|---|
| Heritage tourist | • Visits to castles, palaces, country houses<br>• Archaeological sites<br>• Monuments<br>• Architecture<br>• Museums<br>• Religious sites |
| Arts tourist | • Visits to the theatre<br>• Concerts<br>• Galleries<br>• Festivals, carnivals and events<br>• Literary sites |
| Creative tourist | • Photography<br>• Painting<br>• Pottery<br>• Cookery<br>• Crafts<br>• Language learning |
| Urban cultural tourist | • Historic cities<br>• Regenerated industrial cities<br>• Waterfront developments<br>• Arts and heritage attractions<br>• Shopping<br>• Nightlife |
| Rural cultural tourist | • Village, farm or agro-tourism<br>• Ecomuseums<br>• Cultural landscapes<br>• National parks<br>• Wine trails |
| Indigenous cultural tourist | • Hilltribe, desert or mountain trekking<br>• Visits to cultural centres<br>• Arts and crafts<br>• Cultural performances<br>• Festivals |
| Popular cultural tourist | • Theme parks and themed attractions<br>• Shopping malls<br>• Pop concerts<br>• Sporting events<br>• Media and film sets<br>• Industrial heritage sites<br>• Fashion and design museums |

## Heritage tourism

Heritage tourism is concerned largely with the interpretation and representation of the past. Hence it is a branch of cultural tourism that can be something of a political and ethical minefield. Heritage has become increasingly politicised as recognition has been granted to previously marginalised, minority and ethnic groups. The Western-dominated, Eurocentric approach to the study of history and its interpretation as heritage is no longer acceptable in the postmodern, global environment. Chapter 5 deals with some of these complex and contentious issues in more depth, but a brief summary is presented here of the main discussion points.

The study of history is always disjointed and distorted in some way. The quest for absolute truth and the depiction of 'reality' is ultimately a fruitless quest, as evidence is often hard to come by, and its interpretation is subjective and biased. The so-called 'grand narratives' of the past have often been refuted because of their patriarchal and ethnocentric bias. Instead, the social histories of the working classes, women, ethnic minorities and indigenous groups have gradually become the subject of considerable academic and public interest. The existence of plural histories is being increasingly recognised, although this in itself is problematic, since far more gaps tend to exist in the histories of marginal groups. This may be because such groups were often unable to record their own history or were disinherited, and their heritage displaced or destroyed. This is particularly true of ethnic groups and indigenous peoples.

There has been a growth of interest in such forms of history, and the heritage and museum industries are consequently responding to this development. It is still the case that ethnic and indigenous curators are in a minority, and education and training gaps can be identified. Much of the interpretation of ethnic and indigenous peoples is subsequently carried out by white Westerners, many of whom do not have the knowledge base or the empathy to take on this role effectively.

Tourism has sometimes led to increasing support for minority groups, as it has helped to raise their social and political profile internationally. However, interpretation of heritage is often sensitive and controversial. Some forms of heritage are 'dissonant' to certain groups, such as the heritage of atrocity (e.g. war, massacres, genocide). Care must be taken not to deprive groups for whom such collective events are sensitive of the right to interpret and represent this heritage to others. It may also not always be appropriate to develop tourism in such cases.

The 'globalisation' of heritage has manifested itself in the development of the World Heritage List. In recent years, UNESCO have been moving towards a more inclusive approach to the designation of sites, focusing on their historical and cultural rather than aesthetic value. This means that there is more representation of the intangible heritage of indigenous peoples, for example, or the industrial landscapes of the working classes. Although it could be argued that World Heritage Site inscription affords local communities few benefits in real terms, the initiative at least facilitates and helps fund conservation and encourages tourism development, which can sometimes bring great advantages.

In terms of the management of heritage, a number of issues have come to the fore in recent years, particularly in the 1990s when the concept of sustainability was at its height. Many of the debates relate to the dilemmas that confront heritage sites, for example, maintaining the sensitive balance between conservation, visitor management and community involvement. Others relate to the problems of funding and the extent to which the commercialisation of heritage and museums (e.g. through tourism and retail development) compromise their core function. Overall, heritage tourism is a fascinating area, hence two chapters are devoted specifically to the phenomenon.

## Arts tourism

Arts tourism has perhaps developed more slowly than heritage tourism. This is partly due to a traditional reluctance on the part of both sectors to embrace joint initiatives. The arts sector has often been reluctant to accept the value of tourism and the development or expansion of audiences through tourism. It is often felt that audiences composed largely of tourists would be less appreciative of the art form presented, or that the integrity or authenticity of the performance would be compromised in some way. Of course, there is some truth in this fear, especially as the arts have often struggled financially and been forced to adapt their programming to suit more mainstream audiences.

It could also be argued that the arts are more 'global' than heritage, which tends to be geographically specific and spatially bounded (except perhaps some museum collections). In contrast, the arts can be taken to the people, in the sense that theatre, dance, music and the visual arts travel well in the form of shows, performances and exhibitions. However, it is also true that many people, especially in urban areas, do not have to leave their home town to experience the arts, as the same ballet, opera, play or musical can often be seen there. They do not have to visit the place of origin of the art form to gain access to it.

Access has, of course, become a major issue in the arts, particularly in Western societies where the accumulation of 'cultural capital' or 'cultural competence' is often deemed necessary to understand or appreciate the arts. Attempts are now being made to increase access to the arts and to broaden the profile of audiences. In addition, the increasing cultural diversity and multiculturalism within post-colonial societies has led to the proliferation of new and hybridised art forms, many of which need to be better supported and to have their profile raised.

It will be argued in this book that tourism can bring new, often uninitiated audiences to a wide range of arts events and performances, especially those of ethnic and other minority groups who may be struggling to gain recognition of, or funding for, their art form. There has been considerable pressure on the arts in recent years to adopt a more inclusive and democratic approach to their programming and funding, and to focus on issues relating to access. Tourism arguably creates a channel whereby new art forms can gain support (both financial and moral), and helps to broaden the audience for the arts. In many Western societies, where the arts have traditionally been criticised for elitism, this is perhaps a welcome development.

Chapter 8 focuses in particular on ethnic and minority art forms that reflect the culture, traditions and way of life of a particular group. This includes a discussion of the development of festivals and carnivals, which were traditionally community celebrations and now form part of the heritage as well as the ongoing culture of those communities. The development of cultural tourism can contribute to the flourishing of such events, especially as they tend to be part of the 'local colour' of a destination. Although tourism can sometimes play a role in compromising the authenticity of, or commercialising the event, it is generally regarded as a positive development because of its potentially inclusive and participatory nature.

## Urban cultural tourism

The development of urban cultural tourism, particularly in European cities, has become something of a mass phenomenon, and arguably a serious threat to the future sustainability of a number of historic towns. Whereas cultural tourism was traditionally thought to be a niche form of tourism, the proliferation of short-break holidays has fuelled its rapid, often uncontrolled expansion. It is interesting to note, for example, how quickly some of the emerging urban destinations in Eastern Europe such as Prague and Krakow have become overrun with cultural tourists within the space of ten years. Such issues are discussed in more detail in Chapter 4, as well as in Chapter 6, where issues of sustainability and World Heritage site management are considered in some depth. It should be noted, for example, that many historic towns and heritage cities are not only World Heritage Sites, but they have also been granted European Cultural Capital status at some time or other (see Chapter 4). Although this has afforded them additional protection and funding, it has also led to an increase in visitors due to their enhanced status.

However, this book will focus not only on historic cities that have struggled to maintain the fine balance between conservation and visitor management. Emphasis will also be placed on former industrial cities which have declined, and their attempts to use cultural

**Plate 2.1** Prague circa 1990 – now a mass cultural tourism destination

Source: Author

tourism as a means of diversifying their economy and enhancing their image. Chapter 9 provides a detailed analysis of the concept of cultural regeneration and the role tourism plays within this process. Although such cities cannot compete easily with the world's most popular heritage destinations, they can often position themselves in the short-break market as 'alternative' destinations. Their focus is much more likely to be based on their contemporary cultural tourism product (e.g. popular culture, music, sport, shopping, nightlife), and there is now an identifiable market for destinations that incorporate such attractions. Such destinations are likely to attract more mainstream visitors than traditional cultural tourists, but many of these towns can also afford the visitor a glimpse of their former industrial heritage. The development of industrial heritage tourism in both urban and rural areas is a phenomenon that is discussed in more depth in both Chapters 5 and 9.

Clearly, the majority of cultural cities can offer visitors a wide range of heritage, arts and contemporary cultural attractions. 'World cities' such as London and New York are particularly cosmopolitan and multicultural; hence they can offer the tourist something of the world in microcosm. This includes access to many of the world's major cuisines, music, fashion, dance, sport and so on. Although it is recognised in this book that the globalisation process may be viewed as an unequal and oppressive force, it has also contributed positively to the diversification and hybridisation of cultural forms. Hence, the cultural consumer or cultural tourist can gain access to a broader range of new cultural experiences. It can also mean that traditionally local cultural activities or events become increasingly internationalised, or that global events are brought to a wider range of local audiences through technology, communications, media, travel or tourism. Either way, cross-cultural exchange on a world scale has perhaps never been quite so inclusive and participatory, particularly in the global city.

## Rural cultural tourism

It is recognised in this book that there is an increasing need to focus on the future of rural areas and peripheral regions of the world. Many of the world's most economically and socially marginalised groups live in such areas, and it is necessary to consider how far tourism may be considered to be a positive development option. Chapter 3 focuses on some of these debates, especially the impacts of cultural tourism, and Chapter 7 considers the situation of indigenous and tribal groups. In some cases, the encroachment of cultural tourism into the rural habitats of indigenous peoples has become a destructive or parasitical force; however, this clearly depends on the nature and scale of development. It also depends very much on the degree of fragility of the natural environment, and the attitudes and perceptions of local people towards tourism.

In some cases, tourism may be viewed as a positive force for change or growth within rural areas. Many communities are keen to develop tourism in order to counteract or compensate for the decline of traditional industries, particularly agriculture. The development of arts and crafts tourism, gastronomic tourism or ecomuseums can often help to diversify an economy and provide a supplementary or alternative source of income. These forms of tourism development are discussed in more detail in various chapters throughout the book.

It is also worth noting the development of 'creative' tourism in rural areas as discussed earlier in this chapter. Many tourists are partaking increasingly of 'special interest' holidays, which involve active participation in a cultural activity, such as painting, photography, pottery or cookery. Many of these activities tend to take place in rural areas where the landscape is beautiful (e.g. Tuscany or Provence), or in countries with a popular cuisine (e.g. Thailand or India). Many such tourists enjoy staying with families or on farms in order to enhance the authenticity and rusticity of their experience. Indeed, farm tourism or 'agro-tourism' is becoming increasingly popular, especially in countries that are trying to diversify their tourism product; for example, Greece, Italy, Spain and Portugal. Traditionally, these countries have focused on the development of coastal tourism, but they are trying increasingly to diversify into other forms of tourism, such as rural or cultural tourism. In many cases, especially on Mediterranean islands, a combination of these forms of tourism is offered in the villages and small towns. This can sometimes be problematic in terms of sustainability, since it involves imposing tourism on traditional communities which may be unused, or even hostile to such developments (see Chapter 4 for further discussion of these issues).

Chapter 4 also focuses in some detail on the development of cultural tourism in Europe, and some of the problems of attracting tourism to the regions and rural or peripheral areas. Tourism has traditionally been clustered in urban areas and focused on 'must-see' heritage sites, art galleries and museums. Increasingly, however, the development of trans-European cultural initiatives is helping to open up the regions and to promote the cultures and traditions of under-represented groups. Part of the rationale is also to boost the socio-economic development of such areas, which may have declined following the loss of traditional industries. Hence the regeneration of rural areas is also becoming an important theme.

## Indigenous cultural tourism

Many of the most significant impacts of cultural tourism can clearly be felt in those destinations that are located in remote or fragile landscapes, or which are in developing countries, where the relationship between hosts and guests is an unequal one. Chapter 3 focuses in some detail on the development of cultural tourism in a range of environments, particularly those where tourism has had adverse impacts. Although it was assumed previously that cultural tourism was small-scale and more sensitive to local environments and cultures, its increase has led to its impacts becoming as major as any other form of tourism. Therefore, forms of tourism that originally attracted small numbers of tourists (e.g. ecotourism) have now become more and more mainstream. This might include jungle tours, hill trekking or wildlife tourism, all of which tend to involve contact with local or indigenous people, and often tribal groups. Clearly, the impacts of tourism are likely to be quite significant in such environments.

Chapter 7 focuses on the situation of indigenous and tribal people, including their socio-economic conditions and political status, which have tended to be typical of that of any marginalised or oppressed groups. The colonial legacy in many countries has led to the displacement of indigenous people from their lands, the fragmentation of their culture, and the forced discontinuation of their traditions. This has often led to a certain degree of despondency or despair, especially among groups where whole families and communities have been uprooted and dispersed.

Apart from those relating to land rights, some of the most significant issues relate to the interpretation and representation of indigenous culture and heritage. As discussed in Chapter 5, this can become intensely political, and marginalised groups have a difficult task to make their voices heard, let alone gain the financial wherewithal to fund their own projects or enterprises. Cultural tourism is one way in which indigenous peoples can start to rebuild their communities and to renew their sense of pride in their culture and identity. The very fact that tourists are starting to take an interest in the culture, traditions and lifestyles of indigenous peoples has provided a means of strengthening their position. Although cultural tourism is no panacea, it can be viewed as a positive development option among communities which are favourably disposed towards tourism.

Clearly, a number of important issues need to be considered. Many of these relate to the well-documented phenomenon of community-based tourism, and identifying ways in which local communities can become empowered to develop and manage their own tourism ventures. An interim solution might be provided by joint ownership projects, but ultimately there is a need to increase political and financial support, as well as education and training programmes in order to further empower indigenous communities to manage their own tourism projects. Other issues relate to the interpretation of culture and heritage, which should always be carried out in consultation with indigenous peoples, or by indigenous peoples themselves. This could relate to the interpretation of a museum collection depicting the history of indigenous peoples, for example, or a cultural performance which needs to retain its authenticity. The development of cultural centres in some locations has helped to create a space in which indigenous peoples can represent and express themselves in the way which best befits their community.

## Contemporary cultural tourism

Chapter 1 focused on the development of contemporary, postmodern tourism. In many ways, as discussed earlier in this chapter, it is difficult to establish the boundaries or parameters around cultural tourism if it is defined as a whole way of life. However, calls for access, inclusion and democracy in heritage and the arts have somehow necessitated the broadening of the boundaries and the embracing of all forms of culture and cultural activities. Whereas in the past, 'Culture' (with a capital 'C') was largely regarded as the premise of an elite minority, the concept of multifarious 'cultures' has helped to diversify audiences and to increase access. Many of the most unifying forms of culture are those which are popular and enjoyed by the masses; for example, football and pop music.

In tourism terms, it needs to be recognised that the average tourist now wants to partake of a wide range of activities, which may or may not include traditional forms of heritage or arts tourism. More tourists are travelling than ever before; therefore the industry needs to cater for a broader range of interests and tastes. The majority of tourists are now actively seeking a diverse experience while on holiday, and are just as likely to want to go shopping or to engage in a creative experience as to visit a World Heritage Site. This is a welcome development for non-traditional destinations (e.g. industrial cities or rural areas), as they can now provide tourists with an alternative product and a broader range of activities.

Global developments in technology, media and communications have helped to break down the barriers not just between high and low culture, but also between reality and fiction. Many tourist attractions (e.g. theme parks, leisure complexes and shopping malls) constitute a kind of 'tourist bubble' suspended in time and space, isolated from any real context, and providing the tourist with an idealised environment and experience. Tourism has, of course, always been about the selling of dreams, the creation of fantasies and the perpetuation of myths. The development of simulations in cyber-space and the proliferation of TV travel shows has helped to bring new and potential travel experiences into the living rooms of the public the world over. This has clearly whetted the appetite for travel, and global tourism is consequently growing.

## The advantages and disadvantages of cultural tourism

Although it seems that there is no end to culture, and hence to cultural tourism, these sections have attempted to demonstrate the diverse typologies that can be said to form part of the cultural tourism product. It is by no means a homogeneous phenomenon; therefore its impacts are many and varied. Many of the forms of tourism discussed were originally niche forms of tourism; therefore their impacts were small compared to those of mass coastal tourism, for example. However, the growth of cultural tourism has meant that the impacts have increased in parallel. Although many of these are very positive, care must be taken to monitor the scale and nature of development before destinations become irrevocably damaged. It should not be assumed that the diversification of a tourism product into cultural tourism is necessarily the best development option. For example, in the case of very fragile or remote locations where communities are traditional and close-knit, tourism can become very intrusive and even divisive. It goes without saying that tourism development is not something that should be imposed on communities, but that communities should be allowed to decide for themselves how far tourism is a potentially positive development option. Even then, the size and nature of development should be strictly controlled.

Cultural tourism often appears to be a more environmentally and culturally sensitive form of tourism. This can, of course, be true, but again, as cultural tourism becomes more mainstream and definitions broaden, so then does the profile of cultural tourists. It can no longer be assumed that cultural tourists will have access to better education or superior knowledge; therefore a blanket approach perhaps needs to be taken with regard to the management of sites and the education of tourists. In some cases, local guides can act as 'culture brokers', or informative guides and leaflets can be distributed to tourists by tour operators prior to or on arrival at the destination. Although many destinations attempt to restrict entry or to price out undesirable tourists, it should not be assumed that those tourists with money are likely to be more culturally aware or sensitive. One only has to compare the impacts of the individual backpacker armed with the *Lonely Planet* guide with those of the mass tour group, and this assumption already appears misguided.

Cultural tourism often appears to be an economically desirable prospect for the majority of governments, since it implies an interest in the countries' people, their heritage and traditions, as well as the natural and man-made resources. This can lead to the enhancement of a country's image and the furthering of better international relations, always a priority for governments. However, there is a need for the wealth generated through tourism to be reinvested in the people themselves, rather than being channelled into other economic activities. Only then can the socio-economic and socio-cultural benefits of tourism be maximised and the development of community-based tourism encouraged. Chapter 3 will discuss these issues in more depth.

## Conclusion

The aim of this chapter has been to demonstrate the growing diversity of the cultural tourism product, and to discuss the range of activities that can be said to form part of the product if the concept is defined broadly. Emphasis is also placed on the need to accept broader definitions of the phenomenon in order to embrace concepts of inclusion, democracy and access, which are now so pertinent to debates within cultural studies and cultural politics. It is accepted that cultural tourism in many ways knows no bounds if it is defined simply as a form of tourism that is based on the whole way of life of societies or communities. Therefore, the breakdown of cultural tourism studies into a range of issues based on differing typologies appears to be a logical starting point for the furthering of discussion and debate within the field.

# 3 The impacts of cultural tourism

> There is no end to tourism other than limitless increase. There is no end for the tourist than to visit as many sites as possible.
>
> (Ritzer and Liska, 2000: 155)

## Introduction

This chapter will examine in detail some of the impacts of cultural tourism, in the context of both developed and developing countries. It is often assumed that the negative impacts of tourism development are likely to be more significant in the developing countries of the world, particularly those that view tourism as a panacea for their economic and social problems. However, it is also accepted that cultural tourism can offer something of a boon to developing world economies, as well as making a positive contribution to social and cultural development. It can provide a means of raising the profile of lesser-known destinations and enhancing the standard of living for local people.

These advantages notwithstanding, tourism needs to be managed sensitively and responsibly. In the past, the assumption was made that cultural tourism was a niche form of tourism, attracting small, well-educated and high-spending visitors, hence posing less of a threat to the destination and its indigenous population. However, the growth of international tourism and the diversification of the tourism product have led to an increase in demand for cultural activities, which are becoming an integral part of the visitor experience. The phenomenon of mass cultural tourism is increasingly becoming a cause for concern, whether it is the proliferation of long weekend breaks in the historic cities of Europe, or hilltribe trekking in Southeast Asia.

This chapter will discuss some of the key issues relating to the impacts of cultural tourism in a variety of contexts. These impacts are discussed in terms of their containment and management, placing particular emphasis on the needs and perspectives of local and indigenous communities.

## Tourism and the ultimate quest for paradise

During his visit to an island in the South Pacific, Frank, the protagonist of Jostein Gaarder's novel *Maya* (2000), comments on the apparently inherent human desire to be the first to experience a new place, especially a pristine wilderness. He considers that this is an experience that is possibly surpassed only by becoming the last to see a place before it becomes lost, ruined for all eternity. Smith (1997: 141) notes that some of the postmodern travel modes, such as ecotourism, adventure and wilderness tourism, 'posit a vague awareness of diminishing resources that individuals should *see while they still can*'. We are forever in search of the most idyllic, unspoilt and untouched destination. For example, Lencek and Bosker (1999: 286) describe the search for the perfect beach, and how it has become a symbol of paradise:

> Whatever the beach, it is still possible, in the presence of the timeless wash of waves, the sibilance of sand, and the warm kiss of the sun, to forget the nagging sense of fealty to cash, work, and responsibility. After all is said and done, we still come to the beach to slip through a crack of time into the paradise of self-forgetfulness.

This quest is very well encapsulated in Alex Garland's best-selling novel *The Beach* (1996) (the irony of this being that one of the world's last remaining idyllic beaches in Thailand was partially destroyed while shooting the film of the same name!). The eponymous beach is a jealously guarded secret, but his lead character is more than aware that its future fate is sealed:

> Set up in Bali, Ko Pha-Ngan, Ko Tao, Boracay, and the hordes are bound to follow. There's no way you can keep it out of the Lonely Planet, and once that happens it's countdown to doomsday.

The pessimistic conclusion must be that in time, paradise found is almost invariably paradise lost, especially given that the world is a finite place with finite resources and no place on Earth is now inaccessible. Gaarder's protagonist Frank muses sardonically that:

> Gradually, as the world gets smaller, and the tourist industry develops further niches and sub-niches, I forecast a bright future for necro-tourism: 'See lifeless Lake Baikal!', 'Only a few years before the Maldives are under water' – or: 'You can be the last to see a live tiger!' Examples will be legion, for paradises are getting fewer and fewer, they are both shrinking and being despoiled, but this won't hinder tourism, quite the opposite.
>
> (Gaarder, 2000: 30)

In 1995, television presenter Clive Anderson reported on a number of tourism destinations in a BBC programme that was originally to be called *Trouble in Paradise*. This name was later deemed insensitive, since it implied that the destinations might not be as pristine as they once were. Subsequently named *Our Man In . . .*, the programme examined some of the impacts of international tourism, not all of which were negative. Nevertheless, Anderson (1995: 53) makes a similar point to Gaarder's Frank:

> That 'each man kills the thing he loves' is certainly true of the tourist. We are all looking for the virgin country we can deflower, the unspoiled beach, so that we can be the people to spoil it. The best time to visit any tourist destination is always ten years before you actually get there. Ten years ago the fishing village still had fishermen, and the local bar still had locals. Now, it's full of people like us.

For example, he describes how the people in Goa have often found tourism development to be at odds with existing industries and incompatible with local religions:

> The problems come from a clash between the rich, modern world of tourism and the uncomplicated lifestyle of the folk who have lived in idyllic simplicity for generations. My sympathy was with the Goans struggling to hang on to their paradise.
>
> (Anderson, 1995: 53)

Burns (1999) provides an interesting analysis of the concept of paradise, which is so central to the 'imagineering' of tourist brochures. He cites the principal character of David Lodge's novel *Paradise News*, who ironically notes the way in which the constant repetition of the paradise motif in Hawaii serves to brainwash tourists into thinking that they must surely have arrived there! Second, he quotes the work of Dann (1996a) whose analysis of 5000 images in tourist brochures enabled him to establish different types of paradise, many of which either did not feature local people at all, or depicted them in a stereotyped or contrived way. The conclusion that less than 10 per cent of the images actually depicted tourists and locals together suggests that the notion of paradise is often far removed from the reality of a destination. It also implies that local people are somehow marginal to the experience of a destination, despite its being their home and they the hosts!

The case study (Box 3.1) aims to demonstrate in more detail some of the impacts that tourism has had on the former 'paradise' destination of Hawaii.

## Box 3.1 Paradise lost? A case study of tourism development in Hawaii

Anderson (1995: 205) described Hawaii as the place which 'most closely conforms to the notion of paradise'. Douglas and Douglas (1996) describe how Hawaii has been marketed as 'The Paradise of the Pacific' since the inter-war period, often using the same contrived, commercialised and stereotyped images. However, the mass development of tourism in Hawaii is a well-documented phenomenon. As stated above, David Lodge's novel *Paradise News* gives a comical but perceptive insight into the impacts of tourism in this well-known, over-visited destination. Hawaii has become a popular package tourism destination predominantly for American tourists. Mathieson and Wall (1992: 123) describe how

> Tourists visiting Hawaii's Waikiki's Beach find luxurious high-rise hotels and lavish meals. However they also find crowded beaches, congested streets, water pollution, and contrived images of native life.

Native Hawaiians have been traditionally marginalised in society, and like many indigenous groups they tend to suffer from higher levels of economic deprivation and unemployment. Tourism has done little to remedy these socio-economic problems; on the contrary it has often exacerbated them, displacing local people who can no longer afford the cost of living in the resort areas. Nevertheless, they have little choice but to support tourism because their economic options are now so limited.

Loss of authenticity and exploitation are often inevitable consequences of mass tourism development in such contexts. Cultural activities such as traditional Hawaiian dancing have become a commodified spectacle, and tourists are usually greeted by 'Hula girls' at Honolulu Airport. This is ironic considering that when the Americans originally annexed Hawaii in 1898 they banned the Hula outright, along with the right of local people to speak their language (Trask, 1998). Hula costumes now tend to feature mixed styles and motifs from different Polynesian cultures, and the dances and their performers are frequently eroticised. This of course undermines the spiritual and sacred nature of such indigenous traditions.

Not surprisingly, local responses to tourism development have not been overwhelmingly positive in recent years. Trask (1998: 17) suggests that: 'In principle and practice . . . the tourist industry in Hawaii violates the right of indigenous Hawaiian peoples to self-determination.' De Kadt (1994: 47) cites Pfafflin (1987: 577) who quoted a native Hawaiian as saying:

continued

> We don't want tourism. We don't want you. We don't want to be degraded as servants and dancers. That is cultural prostitution. I don't want to see a single one of you in Hawaii. There are no innocent tourists.

In her article 'Tourism and the prostitution of Hawaiian culture', Trask (2001) comments on 'American imperialism' and the way in which, in the absence of overseas colonies, the Americans have perceived Hawaii as theirs to take. The idea of tourism as a form of imperialism is borne out in the dominance of multinational corporations and landowners who dictate the nature of tourism development. This includes 'mega-resort' complexes and golf-courses. Trask notes that:

- In the past 40 years the ratio of tourists to locals has risen from 2:1 to 30:1.
- Tourism has exacerbated crime rates more than any other factor.
- Tourism is a major catalyst in increasing property prices, inflation and the cost of living for local residents.
- Tourism has been responsible for forcing native Hawaiians to leave their island in search of better economic conditions.

She defines this process as being a form of cultural prostitution, a far cry from the images of paradise depicted in the brochures:

> Thus, Hawaii, like a lovely woman, is there for the taking. Those with only a little money get a brief encounter, those with a lot of money get more. The state and countries will give tax breaks, build infrastructure, and have the governor personally welcome tourists to ensure they keep coming. Just as the pimp regulates prices and guards the commodity of the prostitute, so the state bargains with developers for access to Hawaiian land and culture.
>
> (Trask, 2001: 2)

Minerbi (1996) notes how cultural tourism has been used increasingly as a means of promoting Hawaiian culture abroad, but often simply as a new marketing tool. Ecotourism has been developed since the 1990s in order to diversify Hawaii's tourism product further. Given the significant conflicts between native Hawaiians and the tourism industry, this could be viewed as a positive step forward. However, the emphasis must be on genuine community-based and small-scale initiatives if tourism development is to become more sustainable in the long term.

The quest for paradise is largely an elusive and illusory one, yet human beings still seem to harbour an inherent desire to visit pristine and unspoilt destinations. It is interesting to note that even before we have exhausted the finite resources of this planet, we are eager to explore the next. It is surely only a matter of time before visits to a lunar paradise become commonplace; indeed, the first tourist has already made a trip into space. One small step for space tourism perhaps, but how much of a giant leap for the future of mass tourism?

## Tourism: a new form of imperialism?

The debate about whether tourism is like a new form of imperialism has been prominent since the 1970s (see e.g. Turner and Ash, 1975; Nash, 1977). There is justifiable concern that tourism is, and will remain for the foreseeable future, dominated by Western developed nations, rendering host nations dependent and subservient to its needs. Tourism still flows predominantly from the developed to the developing world. The majority of the world's population, particularly in some of the poorest nations, will never have the chance to venture outside their country, nor perhaps even their home town or village. Hence local people are stationary in both a physical and material sense, and they are often condemned to a life of serving mobile, free-spending Western tourists. The psychological effects of such an unequal relationship are arguably as significant as the socio-economic problems they engender.

Many of the theories relating to the discussion about tourism as a new form of imperialism have their origins in economic development theory. Economists have focused traditionally on core–periphery theory and the growth–dependency relationships between host nations and their Western 'benefactors'. Dependency is viewed as a process whereby the indigenous economy of a developing country becomes reorientated towards serving the needs of exogenous markets (Hall, 1994). Mowforth and Munt (1998) describe how Western capitalist countries have grown as a result of expropriating surpluses from developing countries, which are largely dependent on export-orientated industries (e.g. bananas and coffee). The notion of core–periphery relationships is used within dependency theory to highlight this unequal, often exploitative relationship. Nash (1989) described imperialism as the expansion of a society's interest abroad. Metropolitan centres or cores (usually former imperial nations) exercise their power over peripheral nations or regions of the world. This debate is particularly pertinent to former colonies such as the Caribbean where tourism appears to be reasserting itself as a new form of colonialism. Burns (1999: 157) describes how for dependency theorists:

> development and underdevelopment are two sides of the same coin: surpluses from the exploited countries generated first through mercantilism and later through colonialism, had the combined effect of developing the metropolitan countries and under-developing the peripheral countries.

Mathieson and Wall (1992) suggest that three economic conditions substantiate the claim that tourism is a new form of imperialism or colonialism. These are that:

1. Developing countries grow to depend on tourism as a means of securing revenue.
2. A large proportion of expenditures and profits flow back to foreign investors and high leakages occur.
3. Non-locals are employed in professional and managerial positions.

Hall (1994) suggests that the extent to which tourism will be viewed as a form of imperialism or economic dependency will be determined by the nature of the relationship between the metropolitan centre and the periphery. If it is a hegemonic relationship, its influence will be pervasive in both social and cultural spheres. For example, local people may be unquestioning in their subservience to tourists, since it is viewed as being an inherent, often inherited part of their social and cultural norms.

However, in many ways, it is too simplistic to talk about tourism paralleling the processes of imperialism. As stated by Lanfant (1995: 5):

> the tourist system of action is not a monolithic force. It would be pointless to seize upon it as if it were a hegemonic and imperialistic power perpetuating disguised neo-colonialism. The system is a network of agents: these tap a variety of motivations which are difficult to define and which in concrete situations often contradict each other.

One form of tourism that arguably perpetuates imperialistic relationships of exploitation and dependency is sex tourism (Box 3.2).

The globalisation of tourism has partially exacerbated the relationships of inequality and subservience that are so commonplace in host–guest encounters. It is not simply enough for local people to accept their role as servants, guides or companions to a range of ever-changing tourists. They are also confronted increasingly by the luxurious global products of Western indulgence which remain far from their reach, rather like the thirsty Tantalus in his elusive pool of water. The juxtaposition of local poverty and global wealth is

## Box 3.2 Sex tourism: the exemplification of a dependency relationship

The phenomenon of sex tourism is widespread; yet its covert nature often leads to insidious and uncontrollable growth, which is particularly pervasive in developing regions of the world. Although sex tourism can be variously voluntary or exploitative, commercial or non-commercial, confirming or negating a sense of integrity or self-worth (Ryan, 2000), it can be a considerable cause for concern in some developing countries. Local and indigenous women and men are often rendered subservient to the needs of wealthy, powerful Western tourists. Non-consensual and commercial forms of sex tourism (e.g. those concerning sexual relations between unequal partners in terms of socio-economic status, age, gender or race) are especially rife in Asian countries. As stated by Hall (1992b: 74), 'The sexual relationship between prostitute and client is a mirror image of the dependency of South-East Asian nations on the developed world.' Poverty-stricken villagers are sometimes persuaded to sell their young children to the sex tourism industry as a means of ensuring their family's future survival. It is not uncommon in cities like Bangkok to see young girls with a range of price tags attached being picked out by Western or Asian business men for escort services or sexual favours.

The dependency relationship in the context of sex tourism appears to be based more on wealth and status than on gender. Female sex tourists are not blameless in this process, and it is now becoming more common for Western female tourists to visit destinations such as the Gambia, the Caribbean or India for reasons of sex tourism. However, Sanchez Taylor and O'Connell Davidson (1998) note that adult male prostitutes who cater to demand from female tourists tend to be a lot less vulnerable than women prostitutes who serve a male clientele, as their economic situation is usually less desperate and they are less physically vulnerable.

Sex tourism is arguably part and parcel of the same process of 'othering' whereby local people are depicted (and sold) as being exotic/erotic objects of the tourist gaze. The Western obsession with the cult of youth and beauty is fed through marketing campaigns containing explicit images of young, beautiful, 'exotic' natives. Beddoe (1998) notes how publications such as *Sexual Paradises of the World* (a more sinister use of the paradise motif as discussed earlier) have helped to exacerbate the exploitation of children, young adults and women. By whetting the appetite of the Western tourist for whom money is often no object, such campaigns do little to protect the interests of local people, especially the young, the vulnerable and the poor. Sanchez Taylor (1998) also notes the covert racism in the attitudes of some sex tourists who perpetuate the racist stereotype of the exotic and erotic black woman, especially in such destinations as the Caribbean.

In response, the international campaign ECPAT (End Child Prostitution, Pornography and Trafficking) has done much to raise awareness of a number of important issues. Such campaigns are needed to protect vulnerable groups, but it is a sad fact of life that prostitution is as old as time. So long as tourists have money to spend, and local people have something to sell, the sex tourism phenomenon seems unlikely to decline in the foreseeable future.

particularly stark in some of the world's poorest destinations. For example, in the remote villages of Rajasthan in the middle of the Thar desert in India, foreign trekkers are able to purchase Coke and Fanta from the coolbox-toting villagers, many of whom have barely enough water to survive and have to travel several kilometres a day in order to collect this precious commodity. The same is true of the isolated jungle areas in northern Thailand, where the global brand is omnipresent, yet local villagers are only just subsistent.

## Host–guest relationships and the socio-cultural impacts of tourism

Host–guest relationships have been the subject of much debate and research in a variety of disciplines such as anthropology and ethnography, as well as tourism studies. In many ways, it is difficult to disassociate the impacts of tourism from the broader context of social and cultural development. It is recognised that tourism is only one of a number of global factors that impact upon the traditions and lifestyles of native peoples; hence measurement is difficult and management needs to be viewed holistically.

As discussed in Chapter 2, the profile of the cultural tourist is changing rapidly; therefore it is difficult to generalise about the impacts of cultural tourism. The socio-political context of a destination must be taken into consideration. Equally, many of the models that have been cited so frequently in impact analysis are rendered less useful as tourists proliferate and destinations diversify. However, two of the best-known models are perhaps Butler's 'Lifecycle Model' (1980) and Doxey's 'Irridex' (1975), which complement each other rather well, and, despite their simplicity, retain a certain global relevance (see Table 3.1).

**Plate 3.1** Western tourist interacts with Nubian boys in Aswan, Egypt

Source: Author

**Plate 3.2** Melanie Smith with village women in Thar Desert, Rajasthan, India

Source: Author

**Table 3.1** ***Destination development and local perceptions of tourism***

| *Stages of tourist product life cycle*[a] | *Index of (local) irritation*[b] |
|---|---|
| • **Exploration**: visitor numbers are small, tourist infrastructure is limited, impacts are minimal<br>• **Involvement**: visitor numbers increase, tourist facilities are developed, locals become more involved in tourism<br>• **Development**: the destination becomes a 'resort', arrival of mass tourists, increased external and private sector involvement<br>• **Consolidation and stagnation**: expansion ceases, capacity reached, product quality starts to diminish<br>• **Decline or rejuvenation**: resort either declines further or is revived at a later stage | • **Euphoria**: local enthusiasm for tourism, curiosity, strangers welcomed, mutual feeling of satisfaction for both hosts and guests<br>• **Apathy**: indifference to tourists who become a familiar sight, host–guest relationship less spontaneous and harmonious, tourists targeted for profit-making<br>• **Irritation**: locals unable to cope with the expansion of tourism, often outnumbered by tourists, feelings of exploitation<br>• **Antagonism**: irritations become overt, social unrest, tourists mistreated, targets of crime |

Sources: a: adapted from Butler (1980); b: adapted from Doxey (1975)

Although these models are by no means universal, especially given the complex trajectories of tourism destination development, and differing socio-political and economic contexts, they do allow us to visualise the progression (or, more often, regression) of many global destinations. Many of the earliest tourist destinations such as the Spanish coasts were generally visited by small groups of culturally interested tourists seeking contact with local people. In much the same way as some of the Thai beach resorts are now shifting from bohemian, hippy enclaves to mass rave resorts, the beaches and island resorts of the Mediterranean allowed the scale and nature of tourism to be determined by demand. Many of these destinations now find themselves languishing at the stagnation or decline phase of the resort lifecycle. Their people are often disenchanted, and rarely choose to retain their homes in the same location as the tourists.

In recent years, the regeneration or rejuvenation of destinations has become a more widespread phenomenon. Many destinations have upgraded their product, diversified into new forms of tourism, and are targeting higher spending visitors. In terms of the physical environment, it is perhaps not that difficult to regenerate a resort that has stagnated and declined. However, it is perhaps less easy to win back the goodwill of local residents, many of whom may have already moved out of the tourist destination due to sheer frustration and unhappiness. It could be argued that the economic and environmental impacts of tourism are easier to measure and manage than the socio-cultural impacts, which are often intangible.

It is not uncommon for tourists to be confined to 'enclaves' where contact with local residents is minimal. Chapter 1 discussed the concept of the postmodern 'tourist bubble' in some detail, but there, reference was made largely to themed attractions. In the case of all-inclusive resorts, there are few benefits for the local economy, but in some ways socio-cultural impacts can be managed more easily, as host–guest contacts are minimal and controlled. This relationship allows for little spontaneity, but it is worth questioning how far host–guest relations have ever been truly authentic given the contrived nature and typically short duration of the average holiday. Nevertheless, in many destinations, the problems of all-inclusive holidays have been a major cause for concern (Box 3.3).

## Acculturation, cultural drift and the commodification of culture

It is an inevitable fact of tourism that cultural changes occur primarily to the indigenous society's traditions, customs and values rather than to those of the tourist. In the majority of cases, local people are subjected to a steady stream of changing faces, while tourists vary their destinations with ever-changing frequency. Although tourism may be intermittent and seasonal in some destinations, the constant levels of visitation over time can have a considerable impact on the social and cultural fabric of the host society. In some countries where tourism is largely seasonal (e.g. Greece or Turkey), many local people lead a split life whereby they work in a tourist resort during the summer months, but return home during the winter. This means that the cultural changes that appear to take place within the society constitute a kind of cultural drift rather than acculturation. Anthropologists have been studying acculturation for decades, and it is recognised that tourism is only one of many factors that can lead to permanent cultural change. Mathieson and Wall (1992) differentiate between acculturation and cultural drift, stating that cultural drift is a phenotypic change to the hosts' behaviour which takes place only when they are in contact with tourists, but which may revert back to normal once the tourists leave. Genotypic behaviour is a more permanent phenomenon whereby cultural changes are handed down from one generation to another. This is most likely to occur where tourism is non-seasonal, its influence is strongly pervasive, and local people are favourably disposed towards its development.

There are fears that host culture and identity may be assimilated into the more dominant or pervasive culture of the tourist. The homogenisation of culture is often exacerbated by tourists whose behavioural patterns are sometimes copied by local residents (the

## Box 3.3 The problems of all-inclusive holidays in the Gambia

The Gambia was a former British colony, and it has been suggested that the country has since been recolonised, but this time by tourists. The organisation Gambia Tourism Concern was therefore set up to help establish a more ethical and sustainable form of tourism. Initiatives include the production of a street newspaper called *Concern Magazine* and an in-flight video that is shown on First Choice Air 2000 flights. The video aims to give tourists more awareness of the local people and their environment, and to encourage appropriate and sensitive behaviour. The magazine is sold by local vendors, especially unemployed youngsters on the beaches, who make a percentage on what they sell. It is sold mainly to tourists and has the function of providing information and education.

One of Gambia Tourism Concern's major battles in recent years has been the campaign against all-inclusive holidays, stating that 'The message is loud and clear, the Gambia has more to offer than the dull, uninspiring and secluded worlds that are all-inclusive hotels'. As most of the cost of an all-inclusive holiday is paid for in the tourist's own country, few economic benefits are derived from this form of tourism. Indeed, it is possible for the tourist to remain within the hotel complex for the duration of a holiday, spending nothing in the local economy and having no contact with local people. Even ancillary services will be booked through the hotel. Jeng (2002) states that 'General opinion is that with package holidays, countries like The Gambia get the crumbs from the cake. All-inclusive holidays will deny us even those crumbs.' It seems that the government can do little about the situation at present, as the free market economy allows tour operators and hotel groups total freedom.

Consequently, a campaign has been waged to rally public opinion against all-inclusives instead. It is not simply a question of boosting economic benefits for the destination and its people. The hotels also make very little profit as the price of all-inclusives is so cheap. In addition, it could be argued that many tourists are missing out on some of the Gambia's most memorable attractions and interactions by remaining within the hotel complexes. As suggested by Kempton (2002), there is a need to change the nature of all-inclusives if they are to make a more positive contribution all round:

> Let us start to work for a different kind of 'all-inclusive.' One, which really does include all the players. Which results in an all-embracing relationship, maximum contact between host and guest, experiencing the 'whole' of a country and a culture, its people, its places and its environment in a sustainable way. Let us strive for a tourism industry that provides a mutually beneficial experience for both parties – a tourism industry that is genuinely inclusive rather than exclusive.

Source: Adapted from http://www.gambiatourismconcern.com/http://www.subuk.net/tourism/

'demonstration effect'). This may simply mean that local people feel obliged to learn the language of the tourist in order to converse, but it may also mean the consumption of non-local food or drink, the wearing of non-traditional fashions, and the desire to indulge in the same forms of entertainment as tourists. In non-traditional societies this creates few problems, but in societies which are strictly religious, patriarchal or close-knit, this can impact adversely upon the social fabric. The creation of intra-generational conflicts can become problematic, especially in traditional societies where the younger generations might aspire to Western-style or global living, whereas older generations are keen to protect traditional lifestyles. The role of women can also change rapidly, which can be positive in that it leads to the further emancipation of females within a society. However, in traditional, patriarchal societies this may be the cause of conflict, for example, if women become the main breadwinners within a family.

At some stage, the majority of tourists, even long-term backpackers, tend to crave Western-style amenities. Hence destinations are usually forced to cater to tourists' tastes often supplying fast food, alcohol and brand cigarettes, for example. Not only does this create economic leakages in many cases, but it also threatens the production of local goods, especially if local people develop a preference for Western-style products as well. Conversely, excessive tourist demand for local products can lead to the mass production of traditional goods, which can have the effect of commercialising or commodifying culture. The same is true of traditional events or activities where tourists are keen to spectate or participate. Authenticity becomes a key issue, especially when rituals are performed in isolation from their traditional context. However, 'staged authenticity' in the form of displaced ceremonies, activities and events has become widespread. Although the authenticity of the tourist experience is of some importance, it is more crucial to ensure that local communities feel comfortable with their role as performers and entertainers. This includes the degree to which they are prepared to allow the commodification of their culture for touristic purposes. It should, of course, be understood that some religious or spiritual cultural practices might not be appropriate spectacles for the tourist gaze.

## Measuring the socio-cultural impacts of tourism

It is clearly difficult to measure and monitor the socio-cultural impacts of tourism. Culture is dynamic and changes over time irrespective of tourism development. First, it is difficult to distinguish the impacts of tourism from those of other social or economic developments. Second, few reliable tools exist to measure socio-cultural impacts, which are often intangible. Whereas tools such as economic multiplier analysis or environmental impact assessment can be used to gain quantifiable data, there are few such models that exist for measuring socio-cultural costs and benefits.

However, it should be noted that anthropologists and ethnographers have been researching the phenomenon of cultural change for some time, but it is a complex, time-consuming process. Cooper *et al.* (1998) suggest that a number of techniques can be used to measure socio-cultural impacts of tourism, but that it is usually impossible to filter out other influences. The use of household surveys or questionnaires, focus group interviews, participant observation, or methods such as the Delphi Technique can be used to generate lists of impacts or indicators of social or cultural change. The following list suggests some of the indicators of change that might be used to identify socio-cultural impacts.

- Ratio of tourists to locals.
- Nature of interaction between hosts and guests.
- Local perceptions of tourism.
- Concentration of tourism in certain locations.
- Degree of usage of local products and facilities.
- Extent and nature of local employment.
- Degree of commercialisation of local culture.
- Changes in family relationships and the role of women.
- Demonstration effects.
- Increased social problems (e.g. drug use, alcohol abuse, gambling, prostitution).
- Rises in crime.

(Adapted from Cooper *et al.*, 1998)

## Cultural tourism as a positive development option

Although it is clearly far from being a panacea, cultural tourism can often provide an attractive socio-economic development option for many societies. For example, Lanfant (1995: 3) states that:

> Tourism is often presented as the last chance. Thus, through international tourism, poor regions which have been removed from any focus of activity, closed in on themselves, and condemned to certain death by economists find themselves rediscovered and thrust into the path of development, linked to the international market and propelled onto the world scene.

Tourism can raise the profile of a destination, attracting the interest of investors and visitors alike. It is not simply the case that many countries turn to tourism out of desperation, but because it affords their population a better standard of living. There are a number of benefits that may be derived from tourism provided that it is managed properly. These have, of course, been well documented in tourism literature and might include the creation of employment, the receipt of foreign exchange, the expansion of other economic sectors, and infrastructural developments. In environmental and socio-cultural terms, tourism development can provide a stimulus and funding for conservation, and the preservation of cultural heritage and traditions. Some of the more intangible benefits may include the renewal of cultural pride, the revitalisation of customs and traditions, and opportunities for cross-cultural exchange and integration. Again, these are more difficult to measure, but can be perceptible within local communities.

Inskeep (1994) suggests that a combination of socio-economic impact control measures and socio-cultural programming can be taken to maximise the benefits of tourism development. These are summarised in Table 3.2.

**Table 3.2** ***Impact control measures and programming***

| *Socio-economic impact control measures* | *Socio-cultural programming* |
|---|---|
| • Strengthen linkages between tourism and other economic sectors | • Community and tourist education about tourism and its impacts |
| • Minimise leakages by encouraging use and purchase of local goods and services | • Provide opportunities for cross-cultural exchange and host–guest interaction |
| • Encourage local ownership and management of facilities and services | • Impose visitor codes of conduct where necessary and appropriate |
| • Set a limit on international and non-local investment | • Ensure local community access to cultural facilities |
| • Provide financial incentives for local investment | • Preserve local architectural styles |
| • Maximise local employment opportunities at all levels | • Maintain authenticity of local arts and cultural performances |
| • Provide appropriate training, education and skills development | • Prevent visitation of religious or spiritual sites or ceremonies where appropriate |
| • Encourage and support local business and entrepreneurial development | • Protect and support local cultural production methods |
| • Increase tourist expenditure and the multiplier effects of tourism | • Establish local cultural centres with exhibition and performance space |
| • Develop tourism gradually to counteract problems of inflated prices of land and products | • Market the destination selectively to 'culturally sensitive' tourists |
| | • Limit tourist numbers where necessary through appropriate control measures |

Source: Adapted from Inskeep (1994)

Table 3.2 summarises many of the main issues facing destinations that want to develop tourism in a sensitive and appropriate manner. However, one of the major problems with tourism development, particularly in developing countries, is that the governments of those countries tend to perceive tourism as a 'quick fix' solution to their economic problems. As suggested in the quotation by Lanfant *et al.* (1995), tourism development is often perceived as the last chance for such countries to propel themselves on to the world stage and to compete in the global arena. The temptation is to develop tourism as quickly as possible and to maximise visitor numbers, rather than worrying about the environmental or socio-cultural impacts of doing so. This means that local needs are often bypassed at the expense of the national agenda. Control over tourism development is often relinquished as governments defer to the 'superior' knowledge and expertise of Western consultants and development agencies. International and non-local investment becomes an attractive prospect, especially in poverty-stricken countries where there are unlikely to be many local entrepreneurs who are able to afford the inflated land and property prices.

However, governments need to restrict the extent of outside investment, as otherwise this will lead inevitably to non-local ownership and management of facilities and services and high economic leakages. It is not simply a matter of encouraging local investment where possible; there also needs to be adequate provision of education and training schemes for local people if they are to be employed at all levels of tourism. As discussed above, it is more common for local people to be employed in low-level, often seasonal service jobs where they are subservient to the needs of tourists. Business skills training schemes are also becoming increasingly important, especially if entrepreneurship and small business development are to be encouraged.

Another economic difficulty lies in the inability of many destinations or countries to meet the demands of their tourism industry with their own local products. Although other economic sectors such as fishing or agriculture can be strengthened, supply is unlikely to meet demand if tourism increases rapidly. In the case of small island economies such as the Caribbean or the South Pacific, it might be possible to strengthen inter-island linkages, but there is still the inevitable need to import goods and experience the inevitable economic leakages. As stated above, tourists usually tend to demand Western-style amenities, especially in large global hotel chains and all-inclusive resorts. It is difficult to encourage tourists to spend their money outside the hotel or resort, especially if they are 'psycho-centric' tourists who have no keen sense of adventure or no burning desire to try new things. Although most tourists can be encouraged to shop and buy local handcrafted goods, it is more difficult to insist on the consumption of local cuisine, for example.

The need for local and tourist education is gradually being recognised, as we saw earlier in the case of the Gambia, for example. Tour operators, airlines and Western tourism agencies are being encouraged increasingly to provide information and codes of conduct for visitors. International organisations such as the WTO, Tourism Concern and VSO (Voluntary Service Overseas) have played a pivotal role in raising international awareness of the need for education and information provision. Governments of host destinations are also being encouraged to develop tourism education and training schemes in local schools, colleges and universities, especially in countries where tourism is the main source of income. Codes of conduct may be necessary in areas where the environment or local culture is particularly fragile or sensitive. Certain sites or ceremonies should also remain sacred to local people, who should be free to determine whether the presence of tourists is acceptable to them.

In terms of socio-cultural impact management, it is clear that the relationship between hosts and guests is a difficult one to manage. As discussed above, it is possible to create tourist enclaves or all-inclusive complexes to minimise host–guest tensions. However, it is clear that this arrangement is not at all beneficial in socio-economic terms, and it does not solve the problems of local resentment or misunderstandings; indeed it can exacerbate them. The establishment of appropriate forums for host–guest interaction and the production, exhibition, display or sale of local cultural goods is important. This could be in the form of cultural centres, craft villages or locally run workshops. As long as local needs are prioritised and community control is maximised, this aspect of the host–guest relationship can be managed appropriately and sensitively.

Access to cultural facilities must be guaranteed for locals living within tourist destinations or resorts. This might involve the implementation of a dual pricing system or ensuring free entrance for locals. Where host–guest tensions are rife, there may be a case for adopting different entrance times, but clearly segregation of tourists and locals is not the ideal solution for enhancing host–guest interaction. Access to certain cultural facilities or events may be denied to tourists if local people feel that this is appropriate. Equally, governments may wish to restrict access to certain areas or destinations at any one time, for example, by issuing a limited number of visas.

The final point that should be mentioned briefly is the marketing of the destination. Clearly, many destinations practise selective marketing as a means of ensuring that tourism development remains small-scale and appropriate. It could also be used to ensure that a certain profile of visitor is attracted. However, this is a delicate balance, since the tourists who are the highest spending and will hence benefit the local economy the most may not necessarily be the most culturally sensitive. For example, many backpackers display a greater degree of cultural interest and awareness, but they tend to be low-spending tourists. As discussed earlier in the chapter, access issues are difficult to manage, particularly as tourism is a growth industry and the diversification of the market means that tourists are increasingly keen to visit more remote locations. However, certain forms of tourism, which are more environmentally friendly and culturally sensitive, are being developed in accordance with these changing trends. A good example of this is the ecotourism phenomenon.

## Ecotourism: an ethical alternative?

The year 2002 was declared the International Year of Ecotourism by the UN, whose Commission on Sustainable Development called upon international agencies, governments and the private sector to support activities related to ecotourism.

The Ecotourism Society (1991) defined ecotourism as 'Responsible travel to natural areas which conserves the environment and sustains the well-being of the local people'. The WTO also notes that it includes not only the appreciation of nature but also of traditional cultures present within natural areas. Ecotourism could therefore be defined quite easily as a subset of indigenous cultural tourism as discussed in Chapter 7. Activities might include jungle, rainforest, mountain or desert trekking, hiking, wildlife safaris or bird-watching holidays.

Tourism Concern have already registered their caution over the declaration of the International Year of Ecotourism. It should be recognised that ecotourism is not always synonymous with sustainable tourism just because it has an environmental or cultural focus. As stated by Hall (1994), the footprint of an ecotourist is essentially the same as that of a mass tourist. However, ecotourists tend to be wealthier and higher spending, as stated by Hawkins and Khan (1994: 193):

> Ecotourism involves primarily affluent people travelling from developed countries to developing countries. These ecotourists are from a relatively higher income group with more leisure and more money to spend. They are mostly looking for natural experiences in a pristine environment.

They go on to state that these tourists tend to be environmentally friendly travellers who are keen to see and save natural and cultural treasures, and they are willing to pay more for the privilege hence increasing economic benefits for a location.

However, even where it is small-scale, ecotourism may not always work within the best interests of the local and indigenous population. The degree of local versus external control is a major consideration, although it is accepted that Western influence can often bring significant advancements to indigenous communities. The major problem associated with the phenomenon of ecotourism is the fact that it can open up previously unknown and undeveloped regions to tourism. In the light of the earlier discussion about the expansion of tourism into the remotest locations of the world, this is perhaps an inevitable process. In such cases, ecotourism may be used as a means of moderating and monitoring the scale and nature of tourism in such areas. It can also encourage and educate local people to protect their own habitat better and to focus on the conservation needs of the environment, knowledge that may not be inherent in the local population. Ecotourism development encourages the use of local facilities and amenities, which can help to counteract the problems of economic leakage endemic in other destinations.

However, Mowforth and Munt (1998) suggest that a form of global 'eco-colonialism' has emerged, which is based on the notion that power and control are central to the development of global models of sustainable and environmentally friendly tourism. As stated by Hall (1994), many of the models of conservation found in ecotourism are based on Western concepts. One of the main problems of ecotourism is that many tour operators have tended to abuse the ecotourism label as a marketing ploy, especially in the 1990s, which was predicted to become the 'Decade of Ecotourism' (Smith and Eadington, 1994). As stated by Tourism Concern, ecotourism often tends to fall prey to 'greenwash' marketing, especially as there is no international monitoring system. Like all forms of tourism, ecotourism needs to be managed carefully and sensitively if it is to be sustainable. Table 3.3 suggests some measures that can be taken.

One area where ecotourism may prove to have positive benefits is in Kenya and Tanzania. Although many of the Bushmen in the Central Kalahari in Botswana have been forced off their land because of so-called ecotourism development, the following initiative might well

**Table 3.3** ***Measures to manage ecotourism***

| *Protecting and managing the ecosystem* | *Ensuring local participation* | *Creating economic opportunities for local communities* |
|---|---|---|
| • Maintain appropriate, small-scale, controlled development<br>• Ensure compatibility of development with local surroundings<br>• Aim to conserve energy and recycle materials<br>• Use indigenous labour, materials and expertise<br>• Aim for multiple land-use where possible | • Convey a sense of local ownership, leadership and empowerment<br>• Develop opportunities for local control and management<br>• Create opportunities for group and community-based projects<br>• Respect local heritage, traditions and cultural values<br>• Facilitate host–guest interaction | • Create opportunities for local employment<br>• Ensure local ownership and management of facilities and initiatives<br>• Distribute economic benefits and revenue equitably<br>• Make use of local materials and products to prevent leakages<br>• Channel profits back into conservation, environmental protection and education |

Source: Adapted from Hawkins and Khan (1994)

prove to be successful in rectifying some of the wrongs committed against Masaai tribal people who have been in a similar predicament (Box 3.4).

Clearly, ecotourism is likely to have a successful future if it remains small-scale and local control is maximised. However, the same is true of most forms of tourism; therefore it can be thought of only as a panacea if impacts are carefully and sensitively managed. This

## Box 3.4 Ecotourism development in the Maasai Mara

The Maasai Mara is one of the most popular tourism destinations in Africa, and is one of the best places in the world to view wildlife. It is also home to the indigenous Maasai tribal people, who are semi-nomadic pastoralists whose life centred traditionally on herding cattle. The creation of the Maasai Mara Wildlife Reserve led to local land being taken away from the tribal groups who used the land for grazing their cattle. It is gradually being recognised that it is becoming almost impossible for the Maasai to maintain their traditional culture and lifestyle. Environmental pressures have been enormous, and even local Maasai entrepreneurs have had to compromise the interests of their community. Weaver (1998) describes how the colonial legacy of the park system in Kenya has led to the curtailment of many traditional activities.

Today, wildlife tourism is the leading foreign exchange earner in Kenya and Tanzania, representing 40 per cent of Kenya's foreign exchange. In spite of these enormous economic benefits to the tourism industry and the governments of Kenya and Tanzania, tourism now poses serious threats to the environment, wildlife and culture of the Maasai people. The situation is particularly disconcerting since local tour operators are often completely unregulated. Little of the income from tourism benefits the local people, and many are unable to sustain themselves. Weaver (1998) notes that the majority of rural Kenyans have failed to see any benefits deriving from tourism, especially as local communities tend to be relatively powerless.

However, the creation of cultural Bomas has partly helped Maasai people to benefit financially from tourism and to protect their culture. These are villages created by the Maasai people who want to be able to display their culture and to sell their artefacts to tourists directly without going through the middleman, and they are places where tourists interact with and take pictures of the Maasai people. Cultural Bomas have technically enabled the Maasai to earn some money through tourism directly, although the money has sometimes ended up in the pockets of very few representatives of the Maasai community.

Nevertheless, land-use conflicts are still largely unresolved. Organisations such as Tourism Concern are campaigning on behalf of the Maasai people. MERC (Maasai Environmental Resource Coalition) was founded in 1987 as a non-profit organisation for and by the Maasai people. They are working not only on ecotourism and environmental and cultural protection issues, but also on Maasai indigenous land rights. Landownership is one of the key concerns for the future, along with local empowerment. They hope ultimately to ensure the survival of the Maasai people by preserving their cultural heritage and supporting sustainable socio-economic development within their communities. Some of their recent projects include a community-based ecotourism programme, which was developed for controlled, authorised use of Maasailand in the vicinity of the Maasai Mara Game Reserve with two local tour companies that operate bush walks and safari camps among Maasai communities. Other projects that will generate direct income from tourism through Maasai-controlled centres include selling local crafts and photographic rights, and creating appropriate cross-cultural encounters.

Source: Adapted from Maasai Environmental Resource Coalition (2002)

is largely dependent on the socio-political context of individual countries and destinations, and therefore generalisations are not always useful.

## Conclusion

The aim of this chapter has been to discuss the impacts of tourism in a range of environments, focusing in particular on the measurement and management of socio-economic and socio-cultural impacts. As tourism continues to globalise and diversify, management becomes a more challenging and complex process. Although destinations should be largely learning from the mistakes of the past, it cannot always be assumed that access to such knowledge is available, nor should it be assumed that Western knowledge is always superior. The political, social and cultural diversity of destinations makes it difficult to propose global models and blueprints of excellence or good practice in tourism development. Most countries have a number of restrictions and constraints which may prohibit the development of an entirely sustainable form of tourism. In such cases, these countries should not be denied the opportunity to develop tourism, especially if this is their best economic development option. However, it should be recognised that certain forms of tourism such as cultural tourism and ecotourism can be more environmentally and culturally sustainable, even if they do not yield such high economic benefits through mass development. The values and virtues of such forms of tourism need to be promoted further. The UN's International Year of Ecotourism perhaps provoked a certain degree of scepticism, but at least it engendered important debate, discussion and exchange of ideas on a global scale. Of course, it must be ensured that adequate representation is available from developing countries, and those in which tourism development is currently in its infancy. The future of sustainable tourism development is surely dependent on the sharing of good practice on a global scale. Only then can we hope to protect the few remaining paradises on Earth, and to enhance those that have already succumbed to the simultaneous blessings and blights of tourism.

# 4 European cultural tourism: integration and identity

> Europe is a rich diversity and promotion of this variety on a Europe-wide basis can contribute to the idea of a Europe of the Regions – a Europe unified in diversity, sharing a common but varied heritage.
>
> (ECTARC, 1989: 15)

> A European heritage must accommodate a heterogeneity of place and a multicultural diversity of peoples and cultures.
>
> (Graham *et al.*, 2000: 226)

## Introduction

The aim of this chapter is to discuss the development of European cultural tourism with particular reference to issues relating to integration and identity. The European Union concept of 'celebrating unity in diversity' will be central to the discussion, which will demonstrate the extent to which cultural tourism development can contribute to the celebration of Europe's rich heritage while promoting its diverse contemporary culture. Issues of cultural identity have come to the fore in recent years, partly as a result of political and economic moves towards increasing European integration and unity. For example, the development of European Union, the Maastricht Treaty and the advent of the euro have all contributed to the fostering of better relations between European nations. However, there has also been a concomitant resurgence of nationalism, increased racial tensions and assertion of regional identity in many parts of Europe. A number of European cultural initiatives have been developed to help ease such tensions and to promote mutual understanding. This chapter will attempt to demonstrate the positive contribution cultural tourism can make to this process, focusing in particular on joint European initiatives and projects.

## Issues of cultural integration and identity in Europe

Jean Monnet, one of those responsible for the original development of European Union, declared in his old age that if he had had the chance again, he would have started with culture. Jack Lang, a former French Minister of Culture, later confirmed the importance of culture for the future development of 'Europeanism', stating that we should aim to build a Europe of Culture after having attempted to build an economic and political Europe (Council of Europe, 1994).

There has nevertheless been a certain degree of opposition towards the establishment of a unified Europe (especially in countries such as Britain and Denmark), and it is misleading to consider its people to be one homogeneous, consenting mass. It remains to be seen whether Europe will become truly united in a single political and economic body. For example, there is still some opposition to the euro in a number of European countries, despite its apparent initial success in those that have already adopted it. Countries with a strong national cultural identity may strive to protect their culture from outside influences. The French, for example, fear the impoverishment of their language, popular music and cinema industry as a result of Anglophone influences. Much of this cultural protectionism is arguably more a consequence of globalisation than Europeanisation, but there are still concerns about the political, economic and social aspects of this process.

There has been much debate about the impact of Europeanisation on the culture and identity of the nation state. Sarup (1996) describes how nation states are essentially defined according to a political arrangement of boundaries, as well as a somewhat enforced assertion of common heritage or uniform characteristics. The discourse of the nation state will refer to land, to territory, to myths, to culture. Sarup argues that national identity is an ideology, which helps to maintain and reproduce social power. He states that:

> Discourses of the nation always seem to override class, gender and other social dynamics. By the use of representation, nations recognize and represent the difference between people and make them into a unity. In this way, we are subjected and made into subjects.
>
> (Sarup, 1996: 183)

The inherent human need to belong to such a homogeneous entity may help to fuel nationalist ideologies, which require either the integration or assimilation of extraneous (e.g. immigrant) cultures. However, identities are not static, and they will change according to the strength of different social forces. Hence national identities that are suppressed under political regimes (e.g. Nazism or communism) may eventually reassert themselves in new and different ways. This has certainly been true of many former communist states where the expression of national or regional autonomy and identity has, in some cases, culminated in violent struggles. Immigrants, guest workers or refugees are sometimes forced to integrate or be assimilated into a dominant national culture; thus their identity may become hybridised. This is especially true of diasporic groups where ethnic identity may be fairly strong initially, but which somehow becomes diluted over time as a result of assimilation policies and other social forces. Chapter 7 examines the way in which indigenous identities suppressed under colonial rule have sometimes been revived or revitalised through cultural tourism, and Chapter 8 focuses in more detail on the way in which ethnic identities may be expressed through the arts and culture.

In recent times, the process of Europeanisation appears ironically to have been as much about separation and division as about unification. The 1997 Referenda on Devolution in Britain demonstrated the determination of the Scottish and the Welsh to assert their political and cultural independence from England. The same is true at regional level; for example, Catalonia in Spain, or the Basque Region, not to mention countries with a nationally perceived north–south divide, such as Italy with its Northern League. It is probably true to say that the majority of northern Italians perceive their culture to be more akin to that of their Germanic neighbours than to their Mediterranean cousins in the South. However, interestingly, Bull (1997) describes the Northern League as being more pro-European than many traditional parties and having a belief in a Europe of the Regions rather than a Europe of Nation States.

The problematic reunification of Germany was arguably based as much on a cultural as a political or economic divide. Following the fall of communism in Central and Eastern Europe, there was an almost immediate assertion of regional identity in the former Soviet

Union and Czechoslovakia. Similarly, the Yugoslavian War demonstrated the bitter confirmation of a need for separatism. The political, social and cultural struggles for regional autonomy should not be underestimated. For example, there have been vehement assertions of regional identity in the Basque Country and Corsica. Strubell (1997) warns that such conflicts may become increasingly violent if minority groups continue to be ignored in their quest for recognition and independence.

Regionalism is clearly a strong force which is once again high on the European political agenda. Loughlin (1997) describes how this development is concerned with accelerating the process of European integration. Region can, of course, refer to subnational levels of government or territory as well as to groups of countries. For the purposes of this chapter, regionalism will encompass both definitions. The regions when defined as groups of countries (e.g. Scandinavia, the Mediterranean) are shaped by a number of significant historical, political, geographical, linguistic and climatic factors. Although the nation states within these countries have their own political autonomy and social and cultural variations, they are often defined as regions in the context of tourism studies. Loughlin (1997) identifies four kinds of region:

1 *Economic regions* (e.g. industrialised/deindustrialised; urban/rural).
2 *Historical/ethnic regions* (e.g. societies sharing histories, cultural and linguistic features).
3 *Administrative/planning regions* (e.g. regional entities for policy-making, administration and statistical data gathering).
4 *Political regions* (e.g. possessing democratically elected councils or assemblies).

Paasi (2001) suggests that Europe may be understood in one of three ways: as an experience, an institution and a structure. It is suggested that the experience or feeling of being European has been constrained largely by the existence of the nation state, which dominates spaces of identification. The fostering of a European consciousness and identity would require a more integrative common institutional basis. Paasi notes that the European Union notion of integration has thus far been mainly economic and political rather than cultural. The Europe of the European Union is being increasingly challenged by the emergence of new member states from regions such as Central and Eastern Europe. The representation of Europe as a geographical structure or entity has focused increasingly on sub-European regions. Paasi describes these regions, their 'boundaries' and symbolic meanings as social constructs, expressions of meaning associated with space, representation, democracy and welfare.

The organisation ECTARC (European Centre for Traditional and Regional Cultures) supports and promotes the regional cultures of Europe, placing emphasis on the need for regional decentralisation. Their 1989 *Charter for Cultural Tourism* focused on Western Europe, and in particular on case studies of Wales, Spain and Italy where regionalism is strong. Cultural tourism was perceived as being a means of broadening awareness of Europe's common heritage; increasing co-operation and understanding between countries; strengthening and confirming cultural identities; enabling underdeveloped or declining industrial or rural areas to use cultural tourism as a new source of income; broadening the distribution of tourism in time and space; and contributing to the conservation and enhancement of the built and natural environment. Although such initiatives do not necessarily help to further the social or political cause of the regions, they provide a means of raising the profile of an area and cross-cultural exchange.

Some authors have expressed concern that the recent growth of Eurocentricism has helped to create more dichotomies than ever between north and south, and east and west, and between modernity and tradition (Owusu, 1986). This relates in particular to some of the racial tensions that exist currently in Europe. Mass immigration into European countries following decolonisation and political conflict has resulted in a growing diversification of the population. This has led to a rich diversity of cultures and the emergence of diasporic art forms (for example, see Chapter 8). However, politically, economically and socially, many tensions and conflicts remain unresolved. Some of the worst racial attacks since the Second World War have taken place in recent years, and there is a general concern that the Maastricht Treaty has failed to take into consideration some of the minority groups of Europe. The emergence of a global culture, aided by mass media and technology, has intensified the communication between cultures, peoples and nations throughout the world. However, there is also evidence of increasing fragmentation into nationalistic movements and ethnocentrism. Graham *et al.* (2000: 94) suggest that a European heritage of integration tends to be eclipsed by the Continent's dominant heritage of atrocity, conflict and war:

> the complex array of regional cultural outliers and 'misplaced' political boundaries, accented by distinct dispersed (but mainly urban) minorities subjected to past atrocity or present intolerance, creates a rich medium for dissonance based upon cultural grievances and opposing nationalisms.

Chapter 5 focuses in more detail on issues relating to dissonant heritage, referring in particular to Europe's history of atrocity. Unfortunately, in many European countries, policies of assimilation still assert that people from different cultures and races should adjust to the dominant national or European culture.

The assertion of a cultural identity within such a complex social and political structure is problematic. Paasi (2001) describes the competing discourses that exist on the meaning of 'European identity', and the 5000 or so websites that have been dedicated to this theme. The identity crisis is not simply confined to the many diasporic, dispersed or displaced ethnic or minority communities that have now populated Europe for generations. Post-imperial European societies have also struggled to come to terms with the decline of their empires and their global dominance. The nation state which had reigned supreme for decades is now being apparently subsumed into a unifying Europe. This has been coupled with the globalisation phenomenon which has been dominated mainly by the USA and increasing competition from the Tiger Economies of Asia. The political and economic union of Europe seems to make perfect sense if European nations are to maintain their status as global players. However, culturally this unification is a much more complex and contentious process, especially since the 'decommunisation' of Central and Eastern Europe. Graham *et al.* (2000: 69) describe how many Central and Eastern European countries have rediscovered their national heritage from the pre-communist era, and are even harking back to their formal imperial greatness (e.g. Hungary):

> the wider recent experiences of Central and Eastern Europe serve to perpetuate compelling evidence of the enduring significance of nationalism as the primary mode of identity and national heritage as a principal means of delineating and representing that identity.

In his brilliant history of Europe, Norman Davies (1997: 1136) states that:

> Sooner or later, the European Community in the West and the successor states in the East must redefine their identities, their bounds, and their allegiances. Somehow, at least for a time, a new equilibrium may be found, perhaps in a multilateral framework. . . . Europe is not going to be fully united in the near future. But it has a chance to be less divided than for generations past. If fortune

> smiles, the physical and psychological barriers will be less brutal than at any time in living memory. Europa rides on.

The redefining of identities is one of the most challenging aspects of European integration. In his insightful and humorous analysis of the apparent English identity crisis, Jeremy Paxman (1998: 266) states that:

> The new generation are redefining their own identity, an identity based not on the past but on their own needs. In a world of accelerating communications, shrinking distances, global products and ever-larger trading blocks, the most vital sense of national identity is the individual awareness of the country of the mind. . . . The new nationalism is less likely to be based on flags and anthems. It is modest, individualistic, ironic, solipsistic, concerned as much with cities and regions as with counties and countries. It is based on values that are so deeply embedded in the culture as to be almost unconscious. In an age of decaying nation states it might be the nationalism of the future.

This is an important quotation, since it demonstrates the need for Europeans to think increasingly in terms of the present and not only the past. Identity is something to do with the way we feel rather than how we are defined by others. In an increasingly globalising world, as Europe unifies further, and with the assertion of regional and local identities, the concept of the nation state is perhaps becoming more and more redundant. We need to redefine who we are in the light of these developments, to combat the insular and inward-looking attitudes and prejudices that help to fuel nationalism and racism. The remainder of this chapter will attempt to demonstrate the extent to which cultural tourism can help to further the process of integration and the assertion of regional, local and ethnic identities. The following sections will analyse some of the cultural initiatives in Europe that have contributed to this process.

## European cultural initiatives and joint projects

As early as 1960, the Council of Europe began to take an interest in what it saw as the collective awareness of European cultural high points and their incorporation into the leisure culture. In the past few years, the Council of Europe has begun to realise the importance of cultural tourism and the idea of discovery through travel, both for Europeans and non-Europeans alike. Although the Maastricht Treaty devoted little more than half a page to the subject of culture, there is evidence of a growing desire within the EU for increased trans-European co-operation. This is clearly illustrated by the development of cultural itineraries and networks, many of which are recognising the need to establish a European identity while acknowledging national and regional cultural diversity. In 1964, a Council of Europe working group called *L'Europe Continue* established some objectives which recognised the potential value of cultural tourism and the development of cultural networks. Three main objectives stated in their report were:

1. To raise awareness of European culture through travel.
2. To consider the possibilities of setting up networks for tourism connected with the cultural geography of Europe.
3. To promote the major sites and crossroads of European civilisation as places of interest to tourists.

## Cultural itineraries and themed routes

Since 1987, the Council of Europe has been developing a series of cultural itineraries or themed routes, which often span several countries. These are defined as:

> a route crossing one or two more countries or regions, organised around themes whose historical, artistic or social interest is patently European, either by virtue of the geographical route followed or because of the nature and/or scope of its range and significance. Application of the term 'European' to a route must imply a significance and cultural dimension which is more than merely local. The route must be based on a number of highlights, with places particularly rich in historical associations, which are also representative of European culture as a whole.
>
> (Council of Europe, 1995)

From the outset, three main challenges were established. These focused on making the programme a catalyst for European social cohesion, the establishment of identity, and the extension of cultural tourism to a broader section of society. The themes are generally of historical, artistic or social interest, and the principal motivating factor behind the development of such cultural initiatives appears to have been to encourage the residents of Europe to discover and appreciate their diverse yet common cultural heritage. The Council for Cultural Cooperation established three main objectives for the Cultural Routes programme:

1. To make European citizens aware of a real European cultural identity.
2. To preserve and enhance the European cultural heritage as a means of improving the surroundings in which people live, and as a source of social, economic and cultural development.
3. To accord a special place to cultural tourism among European leisure activities.

The cultural routes were established in order to foster solidarity and tolerance, exchanges between partners from different countries, and the involvement of national and regional institutions, private individuals, groups and voluntary organisations. In addition, such initiatives aimed to lead to an increase in tourism and hence contribute to the economic development and enhancement of the regions concerned, including the creation of employment. The experience that such initiatives hope to afford visitors is an authentic contact with Europe's culture and heritage. At present, the cultural routes programme involves the co-operation of over 2000 partners, and there are over twenty themes (Council of Europe, 2002). These include:

- The Pilgrim Pathways (Santiago de Compostela and the Via Francigena)
- Rural habitat
- Silk and textile routes
- The Monastic Influence routes
- The Celts routes
- Mozart route
- Schickhardt itineraries
- The Vikings routes
- The Hanseatic Cities routes
- The route of parks and gardens
- European Cities of Discoveries route
- Living arts and European identity

- The Phoenician routes
- The Gypsy route
- The route of humanism
- Fortified military architectures in Europe
- The 'Legacy of Al-Andulus' routes
- The Northern Lights route
- Popular festivals and rites in Europe

(www.coe.int, 2002)

The Santiago route was the first itinerary to be developed by the Council of Europe in 1987. The Santiago pilgrim routes form a symbolic route mirroring over a thousand years of European history. Millions of pilgrims have been crossing the Continent since the ninth century on their way to Santiago. The final stretch of the route runs across the south of France and the north of Spain. The experience of travelling the Santiago route is a powerful one, for example, as discussed by the Brazilian author Paolo Coelho (1992) in his book *The Pilgrimage*. The Council of Europe has been ensuring that the routes are identified over Europe, signposting them with a common emblem, and producing promotional publicity material, in particular maps. The Council has also developed a whole cultural programme around them. The programme is linked partly to cultural tourism development, education, conservation and interpretation.

It appears that most of these routes are based on Europe's heritage and the past rather than on its diverse contemporary culture. However, it is clear that there has been some diversification into the culture and heritage of Europe's marginalised or minority communities. For example, The Gypsy route tries to overcome negative stereotypes and prejudices by celebrating Gypsy culture, and the Rural habitat route focuses on the tangible and intangible culture of rural and agricultural communities. Many of the routes have also opened up to include countries and regions from Central and Eastern Europe. These are welcome developments, as they reflect the changing nature of European culture, and support the notions of integration and inclusion.

There is some doubt that such itineraries are, as yet, particularly successful in attracting large numbers of tourists. This fact is due partly to the Council of Europe's insistence that these cultural routes are not a product, and a general refusal to promote them as such. Few routes are as well known as the Santiago route, but this is perhaps a blessing, as over-commercialisation of an initiative that is fundamentally about integration and identity would no doubt be detrimental in the long term to the Council's original objectives.

## Cultural networks

In 1991, the ministers responsible for culture in the EU member states issued a *Resolution on European Cultural Networks*, which encouraged cultural organisations to participate in European-wide co-operation and invited the European Commission to explore the opportunities that networking could offer to member nations. The EU aimed to support those networks that genuinely facilitate trans-frontier co-operation, and are extending their work to new countries especially in Central and Eastern Europe. UNESCO also recognises the importance of cultural networking, and has agreed to support the interconnection between different cultural networks, as well as the promotion of databanks. The aims of many of the cultural networks that have been established in Europe in the past few years are based principally on the exchange of ideas and experiences, and the establishment of joint promotional schemes. They can be used as a means of stimulating east–west exchange, or encouraging joint research and other initiatives that improve information exchange among developed and developing countries, and international organisations. At present, there appear to be over twenty cultural networks in Europe, all of which serve particular functions, according to different priorities. Many of the networks are concerned with cultural and arts policies and funding, research and education. Examples include the *European Forum for Arts and Heritage* which focuses in particular on cultural policy relating to the arts and heritage sectors; *Culturelink* which is a network for research and co-operation in cultural development; the *Foundation of European Carnival Cities* which provides assistance to carnival cities for the management and financing of events; and *CEREC* which offers advice to companies developing pan-European sponsorship strategies, and promotes business support for the arts throughout Europe.

There are also examples of networks that are concerned specifically with cultural tourism, some of which are independent networks, but which aim to further the European themes of integration and cross-cultural exchange. For example, as discussed earlier, ECTARC is concerned with promoting the regions of Europe for cultural tourism, and ATLAS (formerly the European Association of Tourism and Leisure Education and Research, now the Association of Tourism and Leisure Education and Research) undertakes a number of important research projects in the field of cultural tourism, as well as organising conferences which focus on this and related themes. One of the more commercial cultural tourism networks is the *Arts Cities in Europe* network (see Box 4.1).

## Box 4.1 Arts Cities in Europe

The Arts Cities in Europe organisation is an initiative of the Federation of European Cities' Tourist Offices (FECTO). Its principal purpose is to promote culturally motivated tourism and to establish an accessible international centralised booking infrastructure for cultural events, such as exhibitions, festivals, ballets or operas. The initiative aims to support and develop the benefits of the European Union in respect of cultural development, and to establish links between European countries, their peoples and their cultures. Arts Cities highlights the extraordinary cultural range of a large number of European Cities, often redefining small, low-profile cities with underrated potential, and strengthening their image by means of higher profile marketing strategies.

The touristic partner of the Arts Cities in Europe organisation, the Institut für Bildungsreisen (IfB) (GmbH), one of the leading specialist tour operators in Europe, has direct contact with local tourist services, and has extended the choice and range of products available to tour organisers and individual cultural travellers. The IfB works closely with local tourism organisations to set up interesting and attractive packages in combination with special events, such as festivals or exhibitions. There are over 250 participants including major theatres, festivals, opera houses and concert halls, and the bookings cover a wide range of classical and contemporary art forms.

There are currently forty-five cities involved in the initiative in sixteen European countries working together to boost the cultural aspect of tourism, and to promote a better knowledge and understanding of European cultures, traditions and lifestyles. Every year, Arts Cities in Europe issues a comprehensive city handbook which contains details of all the cultural institutions and cities participating in the project. *Arts Cities News* is a magazine which is also distributed on a quarterly basis to end customers, as well as an *Arts Cities Trade Bulletin* which is produced quarterly for the cultural tourism travel trade. Arts Cities in Europe also offers consultation services to cultural institutions, cities, associations and societies on how to promote their events for cultural tourism, and how to create and market a cultural tourism product.

Source: http://www.artcities.de/euns/allaboutus.html

The majority of cultural networks were established in recent years; therefore their success has not always been immediate. Many appear to have encountered similar problems of communication and co-ordination. Language barriers can cause difficulties in terms of communication on a European scale, and many negotiations are based on trust, and political problems can sometimes arise between the various organisations involved in the initiative. Data collection often proves to be problematic, especially given the different ways in which individual countries collect measurable data. There is a genuine shortage of financial resources, and although the European Commission provides some funding, the organisations tend to be non-profit-making, and have to rely on the fact that they should eventually generate their own revenue. There is also a distinct need to ease the burden on popular cultural destinations, by promoting those areas of Europe with underrated potential, and extending the economic benefits of tourism to regional areas and smaller cities. To a certain extent, cultural networking can help to address this problem, by grouping together small and medium-sized towns in order to compete with large cities, by promoting their unique character and attractions.

## The European Cities of Culture initiative

The European Cities of Culture initiative was launched in 1985 as a means of bringing European citizens closer together, simultaneously recognising a common European history and heritage, while celebrating contemporary cultural diversity. Since then, the initiative has become more and more successful and high profile. The motivations for wanting to host the event are many and varied. For example, the event can be used as a means of raising the profile of a city, putting it on the map or enhancing its image; furthering cultural development and providing a forum for artistic expression; developing cultural tourism; or for cultural regeneration purposes.

The Cities of Culture have thus far been designated on an intergovernmental basis, and the European Commission grants a subsidy to the selected city each year. By 1991, the organisers of the European Cities of Culture had created a network which enables the exchange and dissemination of information, especially to the organisers of future events. In 1999, the European Cities of Culture was renamed Cultural Capital of Europe and is now financed through the Culture 2000 Programme. The European Parliament and Council of the European Union established a Community action for the event from 2005 to 2019 on 25 May 1999. They recognised the importance of the event in strengthening local and regional identity and for fostering European integration, as well as providing more tangible social and economic benefits for host cities, including the development of tourism and culture. The stated objective of the newly named Cultural Capital of Europe is 'to highlight the richness and diversity of European cultures and the features they share, as well as to promote greater mutual acquaintance between European citizens' (European Commission, 2002). Cities are required to ensure that they not only promote the culture and heritage of their city, but that they also recognise its place in a European context, and involve cultural activities and artists from other European countries.

Cities now clamour to host the event; indeed, in order to mark the millennium year, nine cities were chosen instead of the usual one. The Commission had so much trouble choosing between the nominated cities during this prestigious year that they were democratic in their approach, much to the chagrin of some of the cities which were keen to retain the glory for themselves. These were Avignon, Bergen, Bologna, Brussels, Kracow, Helsinki, Prague, Reykjavik and Santiago de Compostela. Graham *et al.* (2000) argued that the European aspect of the 2000 initiative was somewhat overshadowed, suggesting that the various cities were mainly keen to capitalise on promoting their own cultural products instead. However, the Commission's criteria are quite specific in their emphasis on the promotion of European connections and cross-cultural exchange; hence a number of joint projects (more than seventy) were organised between the various programmes.

Richards (1999a) examines the changing nature of the Cultural Capital event, suggesting that it increasingly reflects the de-differentiation of culture and economy, and is used predominantly by host cities as a means of gaining competitive advantage. Although the event was designed originally to be purely cultural and to bring people of European member states closer together and to encourage cross-cultural exchange, the need to stimulate the economy and provide a catalyst for regeneration are becoming more prominent imperatives. Richards suggests that the ability of cities to attract cultural tourism will depend not only on their availability of cultural resources, but also on the cultural competence or capital of visitors. The attraction of cultural consumption appears to have become a major incentive for cities wishing to host the Cultural Capital of Europe event, although it can be difficult for new destinations to break into the market. However, as demonstrated in Chapter 9, a number of former industrial cities are managing to attract cultural tourists simply by developing a series of innovative and vibrant new attractions. Although there is some debate about the extent to which this can erode local heritage and sense of place, the development of popular new attractions and the preservation of existing features need not be mutually exclusive.

Table 4.1 shows the list of both past and present cities of culture. It is interesting to note that more than one city has once again been chosen in 2001, 2002 and 2004. However, the Commission plans to revert back to one city per year from 2005, but has chosen one country per year to host the event.

In many ways there appears to be little logic or pattern to the nomination of European cities of culture. Some cities are popular capitals, whereas others are little-known, under-visited regional cities. The European Commission sets out a number of generic criteria

**Plate 4.1** Porto – European City of Culture, 2001

Source: Author

**Table 4.1** ***European cities of culture: past, present and future***

| | | |
|---|---|---|
| 1985: Athens | 1990: Glasgow | 1995: Luxembourg |
| 1986: Florence | 1991: Dublin | 1996: Copenhagen |
| 1987: Amsterdam | 1992: Madrid | 1997: Thessaloniki |
| 1988: Berlin | 1993: Anvers | 1998: Stockholm |
| 1989: Paris | 1994: Lisbon | 1999: Weimar |
| 2000: Avignon, Bergen, Bologna, Brussels, Helsinki, Kracow, Reykjavik, Prague, Santiago de Compostela | | |
| 2001: Porto and Rotterdam | 2003: Graz | |
| 2002: Bruges and Salamanca | 2004: Genova and Lille | |
| 2005: Ireland | 2010: Germany | 2015: Belgium |
| 2006: Netherlands | 2011: Finland | 2016: Spain |
| 2007: Luxembourg | 2012: Portugal | 2017: Denmark |
| 2008: United Kingdom | 2013: France | 2018: Greece |
| 2009: Austria | 2014: Sweden | 2019: Italy |

Source: European Commission, 2002

against which cities are judged, but it could be argued that some cities benefit from the event more than others. For example, those that are already high-profile, heavily visited cities would perhaps be better advised to focus on other initiatives rather than raising their profile further and encouraging more visitors. For example, cities such as Prague, Kracow, Santiago and Bruges are already highly congested cities, and perhaps need to *de*-market themselves. The same is true of Oxford and Canterbury which are bidding for UK 2008. However, if the initiative is viewed as a way of furthering European integration rather than a means of boosting cultural tourism, such nominations are clearly justified. In addition, the funds can always be used for conservation purposes or the advocacy of sustainable development. Regeneration has also become a key theme since Glasgow 1990, and it seems that former industrial cities such as Porto, Rotterdam, Genova and Lille are using the event partly as a means of developing cultural regeneration initiatives. The same appears to be true of the 2008 UK bid (Box 4.2).

## Europe as a 'continuous' destination

Europe has always been what one might describe as a 'continuous destination' both for Europeans and non-Europeans alike. Since the days of the Grand Tour in the seventeenth and eighteenth centuries, tourists have taken advantage of the short distances between countries and the increasingly efficient transport services. The tour schedules of many North American, Australasian and Japanese tourists to Europe incorporate visits to more than one European country at any one time. The fact that Europe is perceived as a single destination is an important development for European tour operators, and it is therefore not surprising that transnational marketing initiatives are becoming increasingly popular. Of course, the visitation of one or more cities per day does little to further the development of sustainable tourism. Environmental impacts are maximised, and economic benefits are few. However, such tours could technically be used as a means of dispersing visitor flows and raising the profile of lesser-known destinations.

Americans, Canadians and Japanese form the largest markets from outside Europe, and they tend to be motivated essentially by culture and heritage. This development has been particularly important for Northern European countries lacking beaches and a warm climate,

## Box 4.2 European Capital of Culture 2008 (UK)

The European Capital of Culture 2008 competition in the UK is being used as an opportunity for UK cities to develop and explore their cultural and creative life. Following on from the perceived success of the only previous UK City of Culture, Glasgow in 1990, the potential of the initiative to help regenerate industrial cities has been recognised. Hence many of the nominated cities are former industrial cities which have embarked on a number of prominent cultural regeneration initiatives in recent years, and hope to use the event as a catalyst for further development.

The nominations for 2008 are Belfast, Birmingham, Bradford, Brighton and Hove, Bristol, Canterbury, Cardiff, Inverness, Liverpool, Milton Keynes, Newcastle and Gateshead, Norwich and Oxford. It was estimated at one stage that over sixty cities wanted to enter a bid! Clearly, Oxford and Canterbury are well-established heritage destinations, as is Norwich to a lesser extent. Many of the other cities (with perhaps the exception of Milton Keynes which is a new town, and Inverness which is surrounded by natural heritage), have been focusing in recent years on large-scale regeneration initiatives. Chapter 9 looks at some examples of these regeneration initiatives in more depth.

Many of these cities are ethnically diverse; hence they have scope to focus on the diverse contemporary culture of their ethnic minorities and immigrant groups, as well as their history and heritage. Bradford in particular is promoting its ethnic culture as part of the bid. In accordance with the Council's criteria, the focus in many of the cities appears to be on community involvement and consultation, social inclusion, access and democracy. There is also a noticeable emphasis in many of the cities on the creative industries, new technology, business development and entrepreneurship. The cities arguably have a hard task in having to prove that they are truly European cities: forward-thinking, but with half an eye on the preservation of their heritage, while promoting their diverse contemporary and ethnic cultures.

In general, the Cultural Capital initiative appears to encourage cities to focus on their past, present and future. This includes the heritage of the city and its communities, its diverse contemporary cultural life, and its future developments (e.g. concerning proposed technological, business or cultural developments). The conferring of the status should not be viewed as a kind of trophy or reward, but as an opportunity to invest in culture and urban renewal. Culture is clearly perceived increasingly as being an excellent catalyst for the development and regeneration of European cities.

as cultural heritage can form the basis of their tourism product. This is certainly true in the case of Britain, and marketing campaigns by organisations such as the BTA often aim to promote the country in its entirety as a cultural or heritage destination, especially to the US market. For example, in 1996, the British Tourist Authority took inspiration from the Arts Cities in Europe initiative and set up a three-year campaign called British Arts Cities. Research indicated that the campaign had been successful in promoting leading British cities as centres of cultural excellence, encouraging extra visitor spending in Britain, and widening the regional and seasonal spread of visitors (BTA, 2002).

Unfortunately, cultural tourism still tends to be concentrated in only a few cities, mainly capital cities, or cities with well-known monuments, museums or galleries. Part of the reason for this stems from the fact that many overseas visitors, such as Americans or Japanese tourists, tend to view Europe as a conglomeration of 'must-see' cultural sites, and rarely have the time or inclination to visit regional and local cultural attractions, especially on a first visit.

There are a number of mainstream and specialist cultural tour operators that organise cultural tours of Europe. The cultural products they offer still tend to be predominantly of a 'highbrow' nature, incorporating such elements as art, music, architecture and archaeology, with tours accompanied by expert guides. Although most tours still appear to be centred around the major cultural cities of Europe, there is evidence to suggest that sustainable tourism is being practised by an increasing number of specialised operators. For example, group size is usually limited; several days are spent in a destination in order to benefit the local economy; public transport is used so as to avoid the congestion caused by tour buses; and visitors are often accommodated in small, family-run hotels.

There is still a wealth of cultural and heritage sites throughout Europe which remain unexplored, and have problems attracting visitors, especially in small towns and rural areas. In commercial terms, heritage sites in remote locations tend to experience considerable difficulties, as the cost of attracting visitors tends to exceed the income generated. For these reasons, the growing number of initiatives promoting cultural circuits and themed routes serve a dual purpose, especially as visitors are rarely prepared to travel extensively to visit one site. They can help to ease the concentration of tourism in single cities, while attracting visitors to smaller, lesser-known sites by linking them with more popular destinations.

## The development of the European cultural tourism product

Over the past few years there has been some concern about the apparent decline in the popularity of Europe as an international tourism destination. This is due partly to the increasing popularity and affordability of long-haul destinations, especially in the Asia-Pacific region, but it is also a result of the decline of many coastal resorts which have been degraded environmentally. In order to retain or even regain their market share of tourists, destinations have been forced in recent years to reconsider the nature of their product, and to focus on features or attractions that are uniquely European, namely their culture and heritage. There has been an increasing trend in both coastal and urban areas towards regeneration or revitalisation of attractions and facilities, as well as the enhancement of image through selective marketing campaigns. Diversification of the product into rural or cultural tourism has also become a common means of attracting a more discerning type of tourist.

In the past few years there has been a noticeable shift in demand away from Mediterranean beach tourism in favour of city-based or rural tourism. However, although cultural tourists have existed theoretically for hundreds of years, it has only been in the past two decades that culture and heritage tourism have been identified as specific tourism markets (Richards, 1996). In the past few years it has generally been recognised that the tourism product can be greatly enhanced by the addition of a cultural element, and this former component of special interest tourism is quickly becoming integrated into the mainstream tourism product rather than being a mere adjunct (Zeppel and Hall, 1992). In Europe, there appears to have been a definite trend towards short breaks and activity holidays, as well as increasingly sophisticated entertainment tastes. In addition, cultural entertainment is becoming an important part of business and conference tourism.

The development of cheap air fares within Europe and the internet revolution have led to a proliferation in short break urban tourism promotions. This includes many of the cities in former Eastern Europe (e.g. Prague, Budapest and Kracow) and some previously under-visited Scandinavian destinations such as Copenhagen, Helsinki and Reykjavik. As discussed, in the past cultural tourism tended to be concentrated in a small number of urban destinations, but the choice of destinations is clearly increasing. However, this is doing little to ease the pressure on existing destinations that are currently suffering from over-visitation and congestion (e.g. Venice, Oxford, Dublin, Amsterdam). It has simply meant that more people are travelling than ever before and visiting a broader range of destinations. Hence cities such as Prague and Kracow already have more visitors than they can comfortably manage. In the urban context it could be argued that a form of mass cultural tourism is developing which is increasingly difficult to control and manage.

## Cultural tourism development in Central and Eastern Europe

The development of tourism in Central and Eastern Europe has been generally well documented; hence only a summary is provided here of cultural tourism development. The emergence of this region as a tourism destination has afforded Western tourists in particular some fascinating and relatively low-cost cultural experiences, especially since the fall of communism. As discussed above, the cities of Eastern Europe are tending to attract large numbers of cultural tourists who are keen to witness their architectural, archaeological and artistic treasures. Some of the most popular cities have so far tended to be Prague, Budapest, Kracow, St Petersburg and Moscow, but increasingly there is an emergence of new urban destinations, for example, cities in the Baltic States of Latvia, Lithuania and Estonia. Some of these new destinations are regional capitals or smaller cities rather than capital cities. It is recognised that currently there may be less interest in some of the arguably less aesthetic capital cities such as Warsaw, Bucharest, Sofia or Belgrade. However, the promotion of smaller towns helps to disperse flows of tourists throughout the region and to raise the profile of traditional and rural cultures. For example, Prague was European City of Culture in 2000, but the Czech Republic made an effort to promote a number of regional towns as part of the initiative (e.g. Kutna Hora, Telc and Cesky Krumlov). Romania is still relatively under-visited, but efforts are being made to promote small cultural towns such as Sibiu, as well as the better-known Bran area in Transylvania with its Dracula Castle.

It is surely only a matter of time before attractive cities such as Ljublijana in Slovenia and Zagreb in Croatia become more popular. Dubvrovnik has always been Croatia's main cultural attraction, and remains so, despite heavy bombardment during the war. This is thanks to major reconstruction efforts. The former Yugoslavian states are perhaps taking longer to establish their sense of identity and autonomy and to rebuild their heritage. Interpretation of culture is still a contentious issue and there has been considerable war damage; therefore many museums still remain closed with their collections in storage.

Most Central and Eastern European destinations are currently attempting to diversify their tourism product. It is interesting to note that cultural or heritage tourism has been one of the first forms of tourism to be developed in this region, although hardly surprising given the curiosity of the outside world about life behind the Iron Curtain. Some destinations such as Bulgaria or former Yugoslavia were already developed predominantly as coastal destinations. Spa tourism was always a popular form of tourism in many Eastern European destinations, especially in the Czech Republic (e.g. Karlovy Vary) and Hungary (e.g. Harkany). Spa tourism may also be found in Slovenia, Bulgaria and Georgia. Rural tourism is being developed further, especially in Hungary where village tourism and gastronomic tours are being promoted. Ski resorts and mountain tourism are also being developed in the Tatra Mountains in Poland and Slovakia, the Carpathians in Romania, and the Julian Alps in Slovenia.

The potential for the diversification of the Central and Eastern European tourism product is considerable, and it is also essential if flows of tourists throughout the region are to be dispersed away from the heavily visited cultural cities. As the infrastructural capacity of new destinations is developed further, under-visited and lesser-known regions are likely to attract more and more tourists. Economically, this will be a boon for relatively deprived countries such as Romania or Albania, and visitors will be afforded an even greater range of exciting and unexpected cultural encounters.

Issues of identity construction are still a major consideration for a number of former Central and Eastern European destinations. This is particularly significant when considering the projection of image through marketing and promotional literature. For example, differentiation through specific activities or unique features may not be sufficient to attract large numbers of cultural tourists. The assertion of new and distinctive identities is an issue that is still highly politicised, and one which is linked to complex social and cultural forces. Although tourism can provide the incentive for identity (re)construction, it can arguably do little to resolve political tensions.

## Cultural tourism development in Southern Europe

It is a well-documented phenomenon that many of the resorts of Southern Europe, particularly in Spain, Portugal and Italy, have experienced the decline or degradation of many of their coastal resorts. These destinations have hence been forced to consider appropriate ways of revitalising such resorts. The techniques used have been many and varied, ranging from the upgrading of the quality of the destination, the development of more environmentally friendly or sustainable forms of tourism, the targeting of new markets, or the diversification of the tourism product into rural tourism, business and conference tourism, sports tourism, and, of course, cultural tourism.

For example, in the early 1990s, the Spanish government made a concerted effort to diversify the Spanish tourism product, focusing in particular on cultural tourism. This included capitalising on three major events in 1992, which were the Olympic Games in Barcelona, the Seville Expo and the European City of Culture initiative in Madrid. The Santiago de Compostela route was revitalised in 1993, and the Silver route was revived which follows the old Roman routes from north to south. A World Heritage Cities Project was also piloted which aimed to establish cultural routes in Avila, Carceres, Salamanca, Santiago de Compostela, Segovia and Toledo (Maiztegui-Onate and Areitio Bertolin, 1996).

Portugal has tried similarly to diversify its product away from the heavy concentration of coastal tourism in the Algarve. Again, cultural events have been used mainly as catalysts for cultural tourism development and regeneration, starting with the European City of Culture event in Lisbon in 1994, the Lisbon Expo in 1998, the European City of Culture in Porto in 2001, and the European Football Championship in Lisbon in 2004. Sintra is promoted for its World Heritage Site status, and cities such as Evora and Coimbra are increasingly attracting cultural tourists. Clearly, the concentration of development has been largely urban-based, but the regional and rural areas of Portugal are being developed increasingly for cultural tourism. For example, the Douro region is being promoted for its gastronomy and wine tourism, and the Alto Minho is starting to develop crafts tourism (Fernandes and Sousa, 1999).

Italy is clearly home to a wealth of cultural treasures, but the government has not always capitalised on its cultural tourism resources as well as it might. The decline of the sun, sea, sand product on the Adriatic coast has not affected Italy's tourism product too adversely, as the Amalfi coast and Ligurian coasts have simply been promoted instead as more upmarket, less congested areas. The cultural cities of Italy are certainly over-visited, especially the 'big three': Rome, Florence and Venice. Italy's current challenge is perhaps to promote the regions better and to disperse flows of tourists away from the main tourist centres. Van der Borg and Costa (1996) note that much of Italy's heritage is to be found outside the traditional city destinations, but that there has not been much growth in the supply of heritage for tourism uses. This is attributed mainly to the tensions between tourism development and conservation. The development of cultural tourism in southern Italy has been a much slower process, and again one which is linked closely to political control. In terms of economic enhancement, the South would clearly benefit from the development of tourism; thus previously under-visited regions such as Calabria and Basilicata are increasingly promoting their natural and cultural resources. Questions about investment in tourism infrastructure need to be raised, but they are often complicated by the existence of illicit, underground forces.

Other Southern European destinations have managed cultural tourism development in different ways. Cultural tourism is seen as one of the potential vehicles for diversifying the beach tourism product in Greece, for example (Kalogeropoulou, 1996). Clearly, Greece's archaeological heritage is dispersed geographically throughout both the mainland and the islands; hence promotion of that heritage is used as part of the diversification process. There is also increasing interest in village tourism, gastronomic tourism and crafts tourism on many of the islands.

Turkey has also capitalised on its rich archaeological heritage in order to diversify its predominantly beach-based product, particularly in the western part (there is, of course, some controversy about the geographical division of Turkey between Europe and Asia). Archaeological sites such as Ephesus and Pergamon attract huge numbers of visitors, as do the hot springs of Pamukkale, not to mention culturally rich Istanbul, the battlefields of Gallipoli and smaller heritage towns such as Edirne.

Although some of the southern French coastal destinations have been somewhat overdeveloped, especially in the Cote d'Azur and Languedoc-Roussillon regions, there has arguably not been the same level of degradation as in other Southern European coastal resorts. In addition, southern France affords the visitor a wealth of cultural experiences. For example, vistors can experience the wonderful Roman remains of Nimes, Arles, Avignon and Orange, or the artistic heritage of the Nice region with painters such as Picasso, Matisse, Chagall and Dufy, or the distinctive culture of the Basque region on the west coast.

## Cultural tourism development on Mediterranean islands

The islands of the Mediterranean offer a rich diversity of landscapes and cultures. Protection of these islands has increasingly become of paramount importance, as their resources are more limited than those of mainland destinations, and the heritage, culture and identity of their people is arguably more fragile. Some islands, especially those in Greece, the Balearics and the Canary Islands, have quite developed tourism infrastructures, and crass over-development has not been uncommon. As a result, many of these destinations are now trying to upgrade or diversify into other forms of tourism. For example, Mallorca is developing more environmentally friendly forms of tourism, and Cyprus is diversifying into village and mountain tourism, including the promotion of gastronomy and traditional festivals. Crete is also increasingly promoting its archaeological heritage, village tourism and arts and crafts tourism in particular.

In many ways, product diversification into cultural tourism is as problematic as coastal tourism development, since it simply disperses the flows of tourists into more environmentally and culturally fragile locations. Sardinia has quite a rugged landscape; hence its tourism industry tends to be based on rural tourism as well as on the more established coastal tourism. People are taking an increasing interest in the Sard culture and language, which could have adverse impacts on the islanders. Although Sardinia has a number of protected heritage coastlines, this protection does not extend to the island's communities. Similarly, Sicily has been something of a hidden treasure for a number of years, although it receives a steady flow of visitors. It has a wealth of archaeological and architectural treasures just waiting to be discovered, and, like Sardinia, a unique culture, dialect and gastronomy. However, heritage conservation issues continue to be a major problem, as does reinvestment into tourism facilities.

There has been some resistance to the development of Europe's islands. The most fascinating example is perhaps Corsica. Richez (1996) describes how local community resistance to tourism development has prevented Corsica from following the same pattern of development typical of the Balearic Islands, for example. This is described partly as a local expression and assertion of cultural identity, but it also has a significant political dimension linked to Corsica's desire for autonomy and independence from France. This has manifested itself in a number of vehement protests.

Malta is one of the most interesting examples of a destination where cultural tourism development may prove to be a detrimental means of diversifying the tourism product. Boissevain (1997) describes how the development of cultural tourism in Malta's towns and villages, particularly Mdina, has impacted adversely upon the local population. Local people feel uncomfortable under the tourist gaze which they feel is obtrusive and invasive, and the town is becoming over-commercialised and reduced to the status of a museum or, worse still, a human zoo. Local hostility towards tourism is consequently growing, especially as there has been little or no consultation over the nature of tourism development. This is perhaps an example of a destination where tourism development should have been confined to the coastal resorts. Indeed, the cultural resources of a destination (however intangible) are every bit as fragile as its environmental resources. If governments fail to realise this, the impacts of cultural tourism are likely to be just as damaging as the impacts of mass coastal tourism.

## Cultural tourism development in Northern Europe

Clearly, many Northern European destinations have relied on cultural tourism as a way of developing their tourism industries. In the absence of a warm climate and good beaches, many such destinations rely on their cultural and heritage tourism to draw in visitors. Clearly business and conference tourism is also perceived as being a lucrative sector, and cultural activities and excursions are increasingly forming part of conference tourism products. Destinations such as the UK have traditionally relied on their cultural tourism attractions, especially when confronted with such rural crises as foot and mouth disease. City break destinations such as Paris, Amsterdam and Bruges have been established for some years, and new destinations emerge all the time with the proliferation of cheap flights in Europe.

Britain has variously promoted its heritage and contemporary culture in recent years, at one point trying to diversify into popular culture with its 'Cool Britannia' campaign. However, once the 'feel good' factor of Tony Blair's election in 1997 had passed, people were less keen to embrace this slogan, and the BTA reverted once more to its promotion of Britain's castles, palaces and traditional heritage sites.

Destinations like Germany have traditionally been tourist-generating countries which do not have to rely on tourism for their economic stability or growth, but increasingly Germany is promoting its cultural cities such as Berlin and Munich. The Rhineland also attracts large numbers of tourists. Belgium is attempting to promote lesser-known cities such as Ghent and Antwerp, as well as the more established Brussels and Bruges. Similarly, tourists are recognising the cultural interest of cities like Rotterdam in the Netherlands which was European City of Culture in 2001. Towns like Delft with its famous pottery and Gouda with its cheese production have also been popular cultural destinations for some time.

## Cultural tourism development in Scandinavia

Traditionally, Scandinavia has perhaps been less visited than many other regions of Europe. Much of this has been to do with the perceived high cost of visiting the region, coupled with the cold climate and long, dark winters, particularly in the northern periphery. However, there has been a growth of interest in both the urban and rural areas of Scandinavia in recent years. This has been prompted partly by the growing affordability of air fares to cities such as Copenhagen, Helsinki and Reykjavik, as well as the development of cruises in the region which tend to incorporate visits to Stockholm and Oslo. Copenhagen was European City of Culture in 1996, Stockholm in 1998, and Helsinki and Bergen in 2000, which has helped to raise the profile of these destinations. Some tourists have started to take an interest in specialist tours to the Arctic to see the Northern Lights, or in Christmas tourism to Santa's Grotto in Finland. Equally, the indigenous Sami people of northern Finland and Lapland are a source of interest for visitors. For example, Miettinen (1999) describes how there is a current move towards strengthening the status of Sami people through textile and crafts heritage tourism. However, it is probably true to say that outside the cities, Scandinavian tourism is perceived to be based more on a rural than a cultural product.

## Cultural tourism development in rural areas of Europe

A number of interesting tourism developments have taken place in some of the rural areas of Europe, many of which are linked closely to cultural tourism. For example, the French concept of the ecomuseum (which is discussed in more detail in Chapter 5) should be recognised as an initiative which has helped to preserve both rural and cultural traditions. It has also been quite a popular means of developing agro-tourism in Northern Europe and Scandinavia. Destinations such as Austria are essentially rural apart from urban cultural tourism destinations like Vienna, Salzburg and Graz. Farm tourism has been a major form of tourism development here, as it can help farmers to supplement their incomes and preserves agricultural life. This form of tourism can, of course, be linked closely to gastronomy, wine tourism and cookery holidays. Rural areas in France, Italy, Portugal, Hungary and Slovenia, among others, have all been recently promoting such forms of tourism. A number of trails or packages are developed which enable the visitor to sample regional cuisine, as well as witnessing traditional production methods.

Similarly, arts and crafts tourism is an interesting development in many rural parts of Europe. The protection of traditional methods of crafts and textile production is of great importance to many rural communities, especially in peripheral areas of Europe. It is clear that the souvenir industry has struggled to meet demand in many destinations, especially in areas where tourists are keen to purchase authentically produced rather than manufactured or mass-produced goods.

An interesting initiative called the EUROTEX project was developed by ATLAS partners and funded by DG XVI of the European Commission in the late 1990s. The aims of the project were to develop crafts and textile tourism in three pilot regions: Crete, the Alto Minho in Portugal, and Lapland. In all of these areas, crafts tourism development was perceived to be an important potential economic development tool, and the projects sought to revive traditional production methods and to provide skills training for local people. It also hoped to foster closer relationships between the piloted regions, and to disseminate information and good practice.

Each of the regions is predominantly rural, and, with the exception perhaps of Crete, under-visited by tourists. The development of crafts tourism in such regions can help to raise the profile of the destinations, address problems of seasonality, and diversify the tourism product. Crafts producers can practise or relearn traditional skills as well as gaining business acumen so that they can run their own small businesses. Clearly, this has both economic and socio-cultural benefits for local communities, especially in areas where agricultural industries have declined and unemployment is high. Instead of local people being forced to move to urban areas to seek work, they can remain within rural areas and make a living while preserving their traditions (Richards, 1999b).

## Conclusion

This chapter has attempted to demonstrate the growing diversification of the European cultural tourism product, and the development of initiatives that can foster integration, inclusion and the strengthening of identities. Although cultural development in Europe has traditionally been deemed less significant than political and economic development, there is a growing recognition that culture is the lynchpin in the creation of a unified Europe. Culture is a contentious concept in that it can be the source of numerous social conflicts and ethnic tensions, but it also has the capacity to bring people together in a variety of forums, networks and initiatives. Cultural tourism is clearly a phenomenon that can variously undermine or strengthen local, regional and European identities, depending on the nature, scale and management of its development. Ultimately, its influence should be a positive one given the increasingly inclusive and democratic nature of its scope.

# 5 Cultural tourism, interpretation and representation

> In the museum of the life world, everyone is a curator of sorts.
>
> (Kirshenblatt-Gimblett, 1998: 259)

## Introduction

This chapter will examine the complex relationship between heritage and cultural tourism, including some of the debates concerning the commodification of history and the 'heritagisation' of the past. There are also many controversial and sensitive issues relating to the ownership, interpretation and representation of heritage, which will be discussed in some detail. These are clearly subject areas which have been debated fairly extensively within heritage literature and museum studies, but this chapter will focus mainly on 'populist' heritage and the heritage of groups that have traditionally been marginalised in historical narratives. The postmodern emphasis on the pluralisation of history and the shift from so-called 'grand narratives' has led to increased interest in, and concern for, the culture and heritage of ethnic, regional and indigenous minorities. There is clearly a need to consider the extent to which such groups are sufficiently empowered to represent themselves and their culture, as well as discussing issues relating to access, democracy and the accumulation of cultural capital. Tunbridge and Ashworth's (1996) concept of 'dissonant' heritage will be of interest here, as well as their discussion of the interpretation of sites, which evoke a certain sensitivity within a collective memory, such as sites of genocide.

## The relationship between history and heritage

It is important to consider a few of the debates surrounding the relationship between history and heritage before discussing some of the more politically sensitive issues relating to heritage interpretation and the representation of heritage for tourists. The debates about the transformation of history into heritage are generally contentious, and consensus has rarely been reached among scholars and academics within the heritage field. However, Lumley (1994: 68) is positive about the impact of such debates on the development of the heritage and museums sectors:

> There are many who work in museums and other cultural institutions associated with heritage for whom a less narrow and insular conception of the national past has undoubted attractions. A greater awareness of the constructed rather than the natural character of heritage can help to loosen the grip of myth. For all its inadequacies, the heritage debate may have contributed to this process.

In terms of definitions, heritage has been associated traditionally with that which is inherited or handed on from one generation to the next. Graham *et al.* (2000) differentiate between the terms 'past', 'history' and 'heritage'. The past is concerned with all that has ever happened, whereas history is the attempts of successive presents to explain selected aspects of the past. Heritage is defined as 'a view from the present, either backward to a past or forward to a future' (p. 2). Essentially, heritage is the contemporary use of the past, including both its interpretation and representation. Increasingly, however, the concept of heritage has become associated with the commercialisation or commodification of the past with the growth of the heritage industry.

Hewison (1987) is notoriously damning of the heritage industry, arguing that heritage is 'bogus history' (p. 144) which 'draws a screen between ourselves and our true pasts' (p. 10). He perceives heritage to be static, fossilising the past, or, worse still, distorting historical facts for the purposes of entertainment. Walsh (1992: 103) is similarly critical of the 'heritagisation' of the past, stating that: 'Heritage sites are constructed as 'time capsules' severed from history . . . they represent a form of historical bricolage, a melting pot for historical memories.' He also argues that 'History as heritage dulls our ability to appreciate the development of people and places through time' (ibid.: 113). This implies that heritage is a false representation of the past, which captures a moment or moments in history, and isolates them from any historical context.

In the postmodern world of 'time–space compression', the idea of synchronicity is becoming more significant than diachrony. Thanks to sophisticated satellite technology, communications and multi-media, we frequently watch history as it happens. Walsh (1992) refers to Baudrillard's (1988) concept of the 'dead point', the point at which history ceases to be real because we can no longer distinguish between truth and fiction. This is the realm of 'hyper-reality' where the past is constantly simulated in a world of mass production and hyper-consumerism. Baudrillard's theories are arguably extreme and somewhat nihilistic, but it is worth considering whether there is an element of truth in his apparent doom-mongering. How far are we losing our sense of the past because of the commodification or 'heritagisation' of history? As stated by Walsh (1992: 149), 'One important characteristic of post-modern heritage has been its unnerving ability to deny historical process, or diachrony. Heritage successfully mediates all our pasts as ephemeral snapshots exploited in the present.'

The concept of nostalgia is often central to discussions about the nature of the heritage industry. As stated by Fowler (1992: 119), 'Nostalgia . . . is one of the most powerful motives for contemporary uses of the past.' Hewison (1987) is very critical of the apparent obsession with nostalgia that dominates people's interest in the past. He suggests that this is true especially of nations where industries have declined (e.g. Britain). He believes that nostalgia is responsible for fuelling the fire of the heritage industry and romanticising and fossilising the past: 'Instead of the miasma of nostalgia we need the fierce spirit of renewal; we must substitute a critical for a closed culture, we need history, not heritage. We must live in the future tense and not the past pluperfect' (1987: 146). Hewison is no doubt referring to Jameson's (1984) notion of the past pluperfect as removing the process of any past actions from the present and creating a kind of 'ahistory'. Lowenthal (1985: 4) suggests that nostalgia has been a boon for the tourist industry: 'If the past is a foreign country, nostalgia has made it the foreign country with the healthiest tourist trade of all.' However, Walsh (1992) contends that the exploration of nostalgia is not necessarily a bad thing as people's emotional attachment to what they remember is of paramount importance. Nevertheless, he also calls for a more critical engagement with the past and its links with, or contingency on, the present. However, Merriman's (1991) survey work suggested that people were actually far from nostalgic or romantic about the past, but realistic about the harshness of living conditions in past times compared to the present.

Many writers have been relatively positive about the development of heritage, arguing that heritage is a means of linking past and present, and enlivening history. Lowenthal (1998) claims that heritage is not 'bad history', it is simply a celebration of the past rather than an inquiry into it. He argues that we do not need a fixed past, but one that fuses with the present: 'History explores and explains pasts grown ever more opaque over time; heritage clarifies pasts so as to infuse them with present purposes' (1998: xv). Kirschenblatt-Gimblett (1998: 7) is even more positive, implying that heritage brings history back to life:

> While it looks old, heritage is actually something new. Heritage is a mode of cultural production in the present that has recourse to the past. Heritage thus defined depends on display to give dying economies and dead sites a second life as exhibitions of themselves.

Schouten (1995: 25) is very positive about the creative potential of heritage interpretation to enliven history, stating that: 'History and historical reality are black boxes: we do not know what they contain, but with an input of imagination and good research, the output can be marvellous. Interpretation is the art that makes history "real."' History has a dynamic quality to it, and will vary according to our perceptions and interpretation of it within the present.

## Heritage interpretation

Tilden (1977: 8) defines 'interpretation' as 'An educational activity, which aims to reveal meanings and relationships through the use of original objects, by first-hand experience, and by illustrative media, rather than simply to communicate factual information.' Use is made of a wide range of tools, which are becoming increasingly high-tech and interactive. Emphasis is often placed on the educational purpose of interpretation, as well as the entertainment function. Light (1995: 139) states that 'At the heart of interpretation is informal education. Interpretation is designed to communicate the significance of heritage places, in a manner appropriate to visitors engaged in leisure activities during their leisure time.' This combination of learning and fun has been referred to by Urry (1990) as 'edutainment', a concept that has become central to the leisure and tourism industries. Uzzell (1989: 3) states cynically that 'Interpretation has been regarded as a novel way of pepping up tired tourist attractions and giving them a value-added component.' This suggests that interpretation is serving to enhance heritage sites and to make them more 'fun' for the visitor. This is perhaps not problematic in itself, but there is a danger that the inherent meaning or nature of the site will be compromised in some way. Graham *et al.* (2000: 3) state that:

> Heritage . . . exists as an economic commodity, which may overlap, conflict with or even deny its cultural role. It is capable of being interpreted differently within any one culture at any one time, as well as between cultures and through time. Heritage fulfils several inherently opposing uses and carries conflicting meanings simultaneously.

This is an interesting and complex quotation which demonstrates the relationship between heritage and time, and heritage and place. Heritage is culturally bound and relative, and it can be interpreted in various ways by different groups at different times. Heritage can also be easily commoditised, especially when it becomes a major component of the tourism product. Schouten (1995) argues that the visitor is looking for an experience rather than the hard facts of historical reality, which can be provided through interpretation. Clearly, the non-expert visitor cannot necessarily construct a meaningful and entertaining experience for him or herself, and is generally reliant on the guide, curator or audio tour for an interpretation of what is being visited.

Tilden (1977: 9) sees the role of interpretation as being a way of encouraging visitors to take a less unquestioning and passive approach to their visit: 'The chief aim of interpretation is not instruction but provocation.' Uzzell (1989: 41) echoes this, and also stresses the need for visitors to engage with and learn from the heritage:

> If interpretation is to be a source of social good then it must recognise the continuity of history and alert us to the future through the past. Interpretation should be interesting, engaging, enjoyable, informative and entertaining. But now and again it has to be shocking, moving and provide a cathartic experience.

It is argued that heritage interpretation should provoke a reaction in its recipients. Kavanagh (1996: 13) states that 'museums are places where memories and histories meet, even collide, and that . . . can be an emotional experience.'

However, some authors have argued that many museums are tending to focus on 'soft' heritage in a deliberate attempt to avoid conflict and controversy (Swarbrooke, 2000). Nevertheless, Tunbridge and Ashworth (1996) have provided an in-depth analysis of the concept of 'dissonant' heritage and the heritage of atrocity, which is by its very nature sensitive and controversial, emotive and sometimes shocking. We will return to these concepts later in this chapter.

It could be argued that heritage interpretation does not necessarily need to be a faithful representation of historical facts and events, especially as history is already open to processes of bias, selectivity and distortion. As stated by Lowenthal (1998: 250), 'Heritage by its very nature *must* depart from verifiable truth.' Gathercole and Lowenthal (1994) discuss in detail the Eurocentric and exclusive nature of historical interpretation, as do Jordan and Weedon (1995). Postmodern interpretations of the past clearly favour plural histories over so-called 'grand narratives'. Historical narratives have traditionally been dominated by white, Western males, hence excluding and marginalising minority, ethnic and gender groups. The heritage industry may well dilute, fossilise or distort history further for the purposes of entertaining the public, but some consolation must be taken from the fact that at least the history of those masses is being increasingly represented, and access to a wider audience is being sought.

## The re-presentation of 'living history'

MacCannell (1976: 78) defines 're-presentation' as being 'an arrangement of objects in a reconstruction of a total situation. . . . Re-presentation aims to provide the viewer with an authentic copy of a total situation.' One example of this is the re-presentation of so-called 'living history'. Fowler (1992: 113) is highly critical of this trend in the development of the heritage industry: 'One of the most dangerous varieties of history, much favoured in educational and entertainment circles, is that called "living history".' He goes on to argue that 'living history' is an impossible concept; therefore any attempt to reproduce it must be fraudulent. Sorenson (1989) is less damning, and writes with enthusiasm about the 'historic theme park'. He differentiates between sites that are based around surviving relics, and sites that are entirely artificial and do not depend on historic associations for their success. He refers to the concept of the 'time machine', 'time warp' or 'time capsule', which transports us to another time and another place for the purposes of education or entertainment. He appears to welcome this broadening of heritage interpretation and a new museology which is based on more than just collections: 'the study of living history and the recording of it should be based upon other criteria than the professional paralleling of the acquisitive preoccupations of the hobbyist' (Sorenson, 1989: 72).

Kirschenblatt-Gimblett (1998: 55) is also fairly positive about the 'living history' phenomenon:

> Live displays, whether re-creations of daily activities or staged as formal performances, . . . create the illusion that the activities you watch are being done rather than represented, a practice that creates the effect of authenticity or realness. The impression is one of unmediated encounter.

However, she goes on to discuss the problems of indigenous cultural performances and live exhibits. These are slightly different issues that we will return to in the latter part of this chapter, and will discuss in more depth in Chapter 7. 'Living history' is generally a re-creation or representation of past events, activities or situations, rather than an interpretation of culture as it is today.

Kirschenblatt-Gimblett (1998) refers to the three different clocks of historical re-creations. There is the stopped clock of the historical moment chosen for re-creation (1627 in the case of Plimouth Plantation), whereby actors are 'unable' to recognise any events after that date in time. Second, although heritage sites technically stop the historical clock by re-creating and constantly re-enacting a moment in time, the heritage clock keeps ticking as time moves forward and the site grows older. Third, there is the tourist clock, the real time of tourists' lives during which they experience the site. Actors sometimes struggle to stay in character when confronted with the real time experiences of the tourist. Actors are challenged to 'forget' the future they know while attempting to re-create an unfamiliar past. It is a challenging experience for all concerned, including the tourist, who may feel slightly unnerved by the apparent authenticity of their encounter with the actor!

## Ownership of and access to heritage

There are key debates within the heritage field surrounding 'ownership' and use of heritage. This includes a focus on cultural politics and the power struggles that seem to govern the interpretation and representation of heritage. Walsh (1992: 79) sees heritage as inherently elitist: 'The public heritage is undoubtedly an extremely narrow and selective concept founded on a dismissal of the richness and variety of what different groups consider to be *their* heritage.' He goes on to be even more critical of the hegemonic and ideological structure, which determines whose heritage is of value:

> The aim of heritage would appear to be to select only that which pleases the sensibilities of a narrow group of people. Those who decide what is worthy of preservation and how it should be preserved, are basically deciding what is worth remembering.

Jacobs (1996: 35) similarly sees heritage as a kind of accolade for which different groups compete: 'Heritage is not in any simple sense the reproduction and imposition of dominant values. It is a dynamic process of creation in which a multiplicity of pasts jostle for the present purpose of being sanctified as heritage.' Lowenthal (1998: 250) refers to the 'adherents of rival heritages [who] simultaneously construct versions that are equally well-grounded (and equally spurious).' Gathercole and Lowenthal (1994: 302) state that:

> The past is everywhere a battleground of rival attachments. In discovering, correcting, elaborating, inverting, and celebrating their histories, competing groups struggle to validate present goals by appealing to continuity with, or inheritance from, ancestral and other precursors. The politics of the past is no trivial academic game; it is an integral part of every peoples' earnest search for a heritage essential to autonomy and identity.

Everybody, not surprisingly, wants their history to be recognised, as well as being given the freedom and resources to interpret it. Merriman (1991) argues that the past belongs to everyone, and that everyone should have access to the past. He perceives museums to be the principal means by which people can gain access to their history; however, he also criticises the failure within the sector to maximise public access to that history. His survey work reveals that the main barriers to access are based on cultural factors (e.g. people's perceptions of museums) rather than structural factors, such as physical access, transport, time, and money. This suggests that people feel excluded from their history, and that there is a need to increase cultural empowerment or cultural capital. Only certain groups (mainly more highly educated people) tend to visit museums, and Merriman argues that museum and heritage visitation is often based more on status affirmation than orientations to the past. He draws on Bourdieu's (1984) notion of the accumulation of cultural capital, which helps to distinguish between social classes. Dimaggio and Useem (1978) demonstrated this in their work on participation in the arts in the USA, stating that participation is often motivated more by reasons of prestige, social display and social status than aesthetic or artistic interest. Walsh (1992: 179) comments on the problems of so-called 'democracy':

> The idea that democratic access is improved through the market is a deception. It assumes a context of democracy in which all members of the public have equal and unrestricted access to both capital and cultural capital, access to the latter being enhanced by greater access to the former. The market is by its very nature undemocratic.

As well as the questions that have been raised about people's access to their own past, it is contentious as to which aspects of the past should be selected for presentation to the public. As stated above, Western imperial societies often tend to idealise or romanticise the past and view it with a sense of nostalgia. Equally, we idealise or glorify the past to reflect society in a positive light. Lowenthal (1985: 332) states that:

> We alter the past to 'improve it' – exaggerating aspects we find successful, virtuous, or beautiful, celebrating what we take pride in, playing down the ignoble, the ugly, the shameful. The memories of most individuals, the annals and monuments of all people highlight supposed glories; relics of failure are seldom saved and rarely memorialised.

Lowenthal highlights one of the most interesting aspects of heritage interpretation, which is our tendency to distort the past and to be selective in our representation of past events. It is common in more 'Eurocentric' interpretations of history to ignore the past of indigenous or ethnic minorities, and to focus solely on the 'glory' of colonial history. Merriman (1991) describes museums in particular as being part of the broader ideological and hegemonic structure which has traditionally determined whose heritage is important or worthy of preservation, and whose is not. Although museums are aiming to move away from this practice, their role and function is perhaps becoming less certain over time.

## The role and function of museums

In the days before international travel became so accessible, museums traditionally served as a form of surrogate travel. Even with increased travel, the advent of the CD-Rom and virtual tours on the Internet has helped to make museums even more accessible internationally without ever leaving the comfort of the PC.

MacCannell (1976: 84) described modern museums as 'anti-historical and un-natural', and it could indeed be argued that many collections do appear to be rather disparate or displaced. Hewison (1987: 84) was unequivocally negative about museums in Britain, describing them as 'symbols of national decline'. This is perhaps overly pessimistic and something of a generalisation, especially given the diverse nature of museums, as illustrated in this comment by Boniface and Fowler (1993: 118):

> Museums are wonderful, frustrating, stimulating, serendipitous, dull as ditchwater and curiously exciting, tunnel-visioned yet potentially visionary. The real magic is that any one of them can be all of those simultaneously.

The character of a museum is determined largely by the nature of its collections. Kirschenblatt-Gimblett (1998: 3) stresses the importance of the presentation of collections:

> Fragmentation is vital to the production of the museum both as a space of posited meaning and as a space of abstraction. Posited meaning derives not from the original context of the fragments but from their juxtaposition in a new context. As a space of abstraction exhibitions do for the life world what the life world cannot do for itself. They bring together specimens and artifacts never found in the same place at the same time and show relationships that cannot otherwise be seen.

However, she later rather cynically cites Washington Matthews (1893) as having said that 'a first-class museum would consist of a series of satisfactory labels with specimens attached' (ibid.: 32), which implies that interpretation is becoming more important and more interesting than the objects themselves! She backs this up later by saying that 'The question is not whether an object is of visual interest, but rather how interest of any kind is created' (ibid.: 78). Walsh (1992: 37) makes a similar statement, but implies that there is a certain bias, not only to the selection of objects but also to their interpretation:

> In a museum display, the object itself is without meaning. Its meaning is conferred by the 'writer,' that is, the curator, the archeologist, the historian, or the visitor who possesses the 'cultural competence' to recognize the conferred meaning given by the 'expert.'

This brings us back to Bourdieu's (1984) idea of cultural capital, and Merriman's (1991) discussion of access. But the question is: How far are museums responding to criticisms of elitism and exclusivity?

Simpson (1996: 265) suggests that museums must become focused increasingly on people rather than collections if they are to evolve. This is not just in terms of access, but also in terms of 'authorship':

> The traditional role of the museum must change, if it is to adapt to the needs of contemporary society, from that of an institution primarily concerned with artifacts and specimens to one which focuses upon people as creators and users of the artifacts in their collection.

Kirschenblatt-Gimblett (1998) believes that museums are experiencing a crisis of identity, and are finding it increasingly difficult to compete with other attractions. They are struggling to shake off their perceived image of being boring, dusty places full of defunct things – effectively 'dead spaces' rather than 'life spaces'. Merriman's (1991) research suggested similarly that people perceived museums to be old-fashioned, musty and dead. He argues that museums should be effectively 'peoples' universities', highlighting the potential of museums as a positive, democratic social force. Walsh (1992: 183) criticises the apparent standardisation of museum services:

> There should not be an emphasis on only one form of representation. A true democracy will offer many and varied forms of museum service. The danger is that we are in fact moving towards an homogenized monopoly of form, which in itself is an attack on democracy.

Evans (1995) describes the growing conflict between the education and tourism function of museums, especially given the growing pressures on museums to raise their own income. Many of the funding and income generation mechanisms are based increasingly on commercial activities such as catering, publishing and retail. The Victoria and Albert Museum in London's 1988 marketing slogan of 'an ace caff with quite a nice museum attached' gave a good indication of this phenomenon. Swarbrooke (2000: 421) takes this discussion further, arguing that whereas in the past museum professionals and the tourism industry 'viewed each other suspiciously across a sea of mutual distrust and ignorance', they were increasingly realising the benefits of collaboration. Tourism has become an invaluable means of generating income. However, he also argues that museums are likely to become the new theme parks for the third millennium if they continue to focus predominantly on entertainment, income generation and marketing at the expense of their traditional core function of being informative and educational. West (1988: 61) makes this point rather cynically in relation to the Ironbridge Gorge Museum in Shropshire, which, he argues, has become increasingly commodified and profit-orientated: 'if pressed to choose I would say that the Pleasure Beach at Blackpool or the amusements at Alton Towers are my idea of a good day out. At least on the big dipper, having paid your money, everyone agrees that the pleasure is in being taken for a ride!'

It is certainly becoming difficult to distinguish between museums and other kinds of visitor or tourist attractions, especially given the advent of interactive technology and multi-media. A good example of this is the highly successful Te Papa Tongarewa Museum in New Zealand (Box 5.1).

It is difficult to criticise such an apparently successful visitor and tourist attraction. It is evident that the experience of visiting a museum is becoming much more of an integrated and interactive experience, but how far it is educational and informative as opposed to just another 'theme park' experience is another matter. Craik (1997: 115) suggests that artificial theme parks are becoming much more successful than most preserved 'themed sites', mainly because they can offer the visitor a more exciting product and integrated experience:

> Compare the product on offer at theme parks with that offered by local history museums: certainly the former recreates impressions of the real while the latter preserves the authentic, but often the latter are immensely disappointing – under-resourced, lacking appeal or diversity, having poor facilities, staffed by enthusiastic but unprofessional volunteers, and so on. As a general rule, such museums are disappointing – and often meaningless – to the tourist.

## Box 5.1 Te Papa Tongarewa Museum

The Te Papa Tongarewa Museum is located in Wellington, New Zealand. The government has made a substantial investment in this new national museum, and hopes to encourage all New Zealanders to visit at least once in their lifetime, as well as promoting it as an international tourist attraction.

It is described in the Mission Statement as:

> a forum for the nation to present, explore, and preserve the heritage of its cultures and knowledge of the natural environment in order to better understand and treasure the past, enrich the present and meet the challenges of the future.

It is a bi-cultural museum where both Maori and Pakeha (white settler) cultures are working together. It includes a living, contemporary *marae*, the first of its kind to be located in a museum. Maori are involved in both the demonstration, interpretation and representation of their culture within the *marae*.

In 1994, Te Papa's Board said that the new national museum would:

- Occupy a central role in the national identity;
- Positively change widely held perceptions of museums;
- Extend the dimension of the museum experience to new levels of engagement and excitement;
- Radically alter the composition and range of the museum audience;
- Focus museum activity more on the primary audience.

Te Papa is described as 'different from any other museum on the planet . . . playful, scholarly, imaginative, educational, interactive, bold'. The museum is high-tech, including two *Time Warp* motion simulation rides *Blastback* and *Future Rush*, and *Golden Days*, an interactive account of key moments in New Zealand's history.

Te Papa has exceeded all expectations, both as a national and international attraction. By 30 June 1999, the museum had received over 2.5 million visitors since its opening in February 1998. It is now firmly established as New Zealand's cultural flagship and leading visitor attraction. It has attracted a significant number of New Zealanders living outside Wellington, as well as large numbers of international visitors. Te Papa has also succeeded in expanding the core audience for museums. Typically, Te Papa's visitors reflect the diversity of New Zealand's population to a much greater extent than has been traditional for museums.

Source: Adapted from Te Papa Tongarewa Museum (2001)

Urry (1990) is much more positive about the changing nature of museums, especially the proliferation of plural histories (e.g. social, feminist, ethnic, industrial, populist) that are being increasingly represented. Some examples of this will be discussed later in the chapter. He argues that museum displays have become much less 'auratic', and he is largely positive about the replacement of 'dead' museums with 'living' ones. However, it could be argued that museums have not yet gone far enough in their representation of plural histories. As stated by Porter (1988: 104), 'museums have been slow to take up issues such as racism, class bias, and sex discrimination, either as employing institutions, or as a medium which propagates a particular and pervasive brand of history.' For example, Porter supports Horne's (1984) contention that traditionally museums have been patriarchal and that representations of women tend to be stereotyped. Carnegie (1996) is critical of the representation of women as mere 'home-makers', rather than being recognised for the roles they played in key industries. However, Simpson (1996: 5) is largely positive about the future of museums in terms of the adoption of alternative perspectives:

> Museums are changing in many ways: their image as dusty, stuffy, boring and intimidating storehouses is slowly giving way to recognition that museums can be authoritative without being definitive; inclusive rather than exclusive; exciting, lively and entertaining while still being both scholarly and educational.

Nevertheless, Swarbrooke (2000) criticises museums for focusing increasingly on 'soft' history in an obvious attempt to avoid controversy and conflict. Urry (1990) also suggested that our emphasis on 'artefactual' history ignores or trivialises a whole variety of social experiences, such as war, disease, hunger, exploitation. Similarly, Kavanagh (1996: xiii) states that: 'In the acceptance of alternative narratives is the recognition that if history in the museum is to be valid intellectually and worthwhile socially, it has to find ways of admitting to and dealing with the darker side of ourselves and our pasts.' This may well be true; this is therefore perhaps an interesting point at which to consider some of the more 'dissonant' and 'darker' forms of heritage that may or may not be represented adequately in our museums and heritage sites.

## Dissonant heritage and 'dark' tourism

Graham *et al.* (2000: 93) define 'dissonant' heritage as:

> the mismatch between heritage and people, in space and time. It is caused by movements or other changes in heritage and by migration or other changes in people, transformations which characteristically involve how heritage is perceived and what value systems are filtering those perceptions.

Dissonance basically refers to the discordance or lack of agreement and consistency as to the meaning of heritage, who 'owns' it, and who should be allowed to interpret it, and in what ways.

As mentioned earlier in this chapter, the process of heritage interpretation is problematic, as it is difficult to offer the visitor a truly objective depiction of historical reality. It usually requires those heritage site managers and museum curators to select and display selected aspects of the heritage, often of a group to which they may or may not belong. If the heritage managers concerned are not members of the group whose heritage is being interpreted, this raises the question of ownership. To whom does the heritage belong, and who should have the right to interpret and present it to others? This may not be a problem in cases where the particular group can no longer claim ownership (i.e. they are an ancient civilisation, or there are no remaining survivors). As Blakey (1994: 39) says, 'Archaeologists speak for a past that cannot represent itself . . . archaeological views of the past are reshaped by changing cultural biases'; however, the feeling of 'disinheritance' may become a crucial issue if a group feel that their rights to their heritage have been usurped by others.

A good example of disinheritance might be the strong sentiments of the Jewish community in Poland, where it was felt that the Polish communists had provided a distorted interpretation of the Holocaust, especially at the site of the concentration camp in Auschwitz. The Jews felt 'disinherited' because, in their view, the Polish had represented the events as being a national or a European tragedy rather than a Jewish one (historical evidence shows that 90 per cent of the victims who died in Auschwitz were Jewish). The

**Plate 5.1** Auschwitz concentration camp: a dissonant site requiring sensitive interpretation

Source: Author

visitor to Auschwitz could view a number of exhibitions which were devoted to the plight of different nationalities, including other minority groups, such as gypsies or homosexuals. Although it was important to represent each of these groups, it arguably distorted the scale of the Jewish tragedy. This situation does raise another important question about representation and the perspective of history offered to visitors. Whose experience or perception of events should be viewed as the most valuable or the most important, and how should it be represented? Should the perspective of the 'victim' always be given priority (in this case the Jewish community during the Holocaust), or is the perspective of the 'observers' (in this case, the non-Jewish Polish community) or the 'perpetrators' (here, the Nazis) equally as important to the 'true' representation of historical events (Tunbridge and Ashworth, 1996)? In theory, if we hope to depict historical events as objectively as possible, then all three perspectives should be given equal priority, and would be included in a display or exhibition of those events. In such a case, however, the potential offence or upset that might be caused would no doubt lead to a very different representation of events.

Just as the interpretation of heritage usually disinherits someone in some way, there are always aspects of the past which will be dissonant or distasteful to certain individuals or communities. Much of this will depend very much on individual sensibilities. Not every tourist will want to make the pilgrimage to a concentration camp such as Auschwitz, or to First World War trenches in France or Belgium, or to a site of genocide such as the 'Killing Fields' in Cambodia. Those who do usually have their own particular motivations for doing so. Some visitors may have experienced a personal loss; others feel a sense of collective tragedy, especially if they are part of a certain community; some visit for educational purposes; some may just have a morbid fascination with the darker side of human nature, arguably a dubious motivation, but one which seems to be common to many visitors to such sites.

What are the implications for heritage site managers of this kind of heritage of atrocity, and to what extent is it acceptable to interpret such sites for the purposes of tourism? We are not referring here to tourism attractions which are purpose-built for leisure and entertainment (e.g. chambers of horrors, museums of torture and so on). Many sites of atrocity, such as concentration camps, massacre sites, battlefields or cemeteries serve as memorials to historical events and are visited as such, often for educational purposes. Within Europe, citizens are generally encouraged to learn about their collective history, however distressing it may be, whereas in other societies a more selective version of history may be taught.

Visits to military heritage sites, such as battlefields, war museums, battleships and so on are also becoming more and more popular, especially in Europe. However, demand for this kind of tourism is also growing internationally. For example, many visitors to Thailand now visit the River Kwae bridge and Death Railway, and Vietnam affords the visitor a plethora of war-related 'attractions' such as the Cu Chi tunnels and Ho Chi Minh Trail.

It should be pointed out, however, that our perceptions of historical events and our approach to heritage interpretation are affected greatly by temporal and spatial factors. It is interesting that we are quite happy to create entertainment through battle re-enactments of the Battle of Hastings, whereas it would be considered extremely insensitive and distasteful to re-enact the trench warfare of the First World War for the purposes of entertainment. While veterans of this terrible war are still alive, and the events are very prominent in the minds of subsequent generations, we cannot feel detached from the historical reality of the situation. Equally, the closer to home an historical event seems to be, the more likely we are to be sensitive to its interpretation. The historical events which are being depicted in the museums and exhibitions of Europe are still very prominent in the European collective memory, and people are sensitive to their interpretation in a way that a non-European might not be. Tourists are presented with a certain interpretation of an historical reality. In many cases this may be their only visit to a site, and their impressions of that site and their understanding of a country or region's heritage will be influenced largely by the nature of the interpretation of the site. It should therefore be managed carefully and sensitively.

Clearly, in some cases, there may be reasons why a government or political leader wants to distort the visitor's perception of the country, and will therefore censor what tourists are allowed to see (this is normally the case with a totalitarian regime, e.g. Myanmar or North Korea). Museums or heritage sites may be used as instruments of political propaganda, or in the furthering of political ideologies.

The case study given in Box 5.2 offers an interesting and controversial example of a museum that has attempted to depict the events of the Holocaust in a graphic and disturbing way.

## Ethnic, indigenous and minority heritage

The notion of disinheritance should perhaps be discussed in more detail, as we have so far talked about the concept of a 'collective' history or heritage without really defining what is meant by this. To what extent does the interpretation of national heritage marginalise regional, ethnic or indigenous minority groups? As stated by Tunbridge and Ashworth (1996: 29):

> The shaping of any heritage product is by definition prone to disinherit non-participating social, ethnic or regional groups, as their distinctive historical experiences may be discounted, marginalised, distorted or ignored. This, it has

## Box 5.2 The U.S. Holocaust Memorial Museum, Washington, DC

The U.S. Holocaust Museum was opened in 1993, and visitor levels have since been in excess of two million per annum. The institution combines memorial with museum and was commissioned by an act of government. The primary purpose of the museum has been defined as educational as well as being a symbolic reminder of the darker side of human nature. Interestingly, another of its main aims is to encourage the American people to think critically about their own political culture and to foster more ethical behaviour in society.

The museum's permanent exhibition *The Holocaust* spans three floors of the museum building. It presents a narrative history using more than 900 artefacts, seventy video monitors and four theatres that include historic film footage and eyewitness testimonies. This includes letters, photographs, uniforms, and a rail car used to take Jewish and other prisoners to the death camps, not to mention the realia of human hair, glasses, shoes and clothes. Visitors are able to hear taped interviews with Holocaust survivors, as well as leafing through newspaper reports and watching video clips.

The exhibition is divided into three parts: 'Nazi Assault', 'Final Solution' and 'Last Chapter'. The narrative begins with images of death and destruction as witnessed by American soldiers during the liberation of Nazi concentration camps in 1945. This is in order to orient visitors who may have no knowledge of history.

Visitors gain a card on entry featuring the identity of a Jewish citizen of Hitler's Germany matching their own age and gender, whose 'fate' can be traced periodically using computerised stations throughout the museum.

Many visitors have found the exhibition highly emotive, disturbing and sometimes distasteful. Some commentators are perturbed by the way in which the citizens' identity cards are discarded in a litter bin outside the museum. Gourevitch (1999: 3) is horrified by the grisly and almost voyeuristic nature of the exhibition, which ironically has an almost anaesthetic effect after prolonged exposure to the graphic scenes. He concludes by saying:

> I cannot comprehend how anyone can enthusiastically present this constant recycling of slaughter, either as a memorial to those whose deaths are exposed or as an edifying spectacle for the millions of visitors a year who will be exposed to them. Didn't these people suffer enough the first time their lives were taken from them?

However, the museum is necessarily controversial, and it arguably makes a valiant attempt to interpret an event which is almost beyond human comprehension, let alone interpretation. As stated by Lennon and Foley (1999: 50):

> The enormity of the systematic destruction of the Jewish people is beyond comprehension, interpretation, and explanation. Language, images, and art are inadequate in this area and the scope of the subject will inevitably remain difficult to comprehend.

Sources: Adapted from U.S. Holocaust Memorial Museum (2001); Gourevitch (1999); and Lennon and Foley (1999)

> been argued, is an innate potentiality and a direct consequence of the selectivity built into the concept of heritage. Choice from a wide range of pasts implies that some pasts are not selected, as history is to a greater or lesser extent hijacked by one group or another for one purpose or another.

Following the First World War, the fall of the empires in Europe meant a shift from imperial to national heritage, meaning that minority heritages were often seen as a threat to national integrity, so they were either ignored, discouraged, disinherited or removed. Gathercole and Lowenthal (1994) suggest that indigenous and ethnic minorities tend to cherish their monuments and sites as being bastions of community identity, especially as they have usually been forced to relinquish their land, religion, language and autonomy under colonial rule. Clearly, many groups are not able to represent themselves because of lack of funds to establish a museum or exhibition, or to interpret a heritage site. Coupled with the frequent destruction of non-European artefacts because of their apparent 'worthlessness', and the displacement of objects from their native lands, indigenous minorities rarely have the political leverage and economic wherewithal to reclaim and represent their heritage.

There are often arrogant assumptions that indigenous peoples need to be educated about their own cultural background and history. Worse still, there were frequently Eurocentric and elitist historians who believed that indigenous cultures were not worthy of representation. For example, the eminent historian Hugh Trevor-Roper (1965: 11) declared that: 'Perhaps in the future, there will be some African history to teach. But at the present, there is none, or very little. . . . The history of the world, for the last five centuries, in so far as it has been significant, has been European history.' In addition to being deeply patronising, this comment is indicative of an intensely subjective and selective approach to the study of history which has dominated for centuries. As Gathercole and Lowenthal (1994) so brilliantly outline in *The Politics of the Past*, there is a heritage of Eurocentricity which has suppressed and undermined the history of colonised communities, a history which clearly pre-dates that of the colonising nation. It is not uncommon to hear people declare that Australia, New Zealand or the USA have little or no history, and few heritage sites compared to European countries, discounting, of course 40,000 years of Aboriginal, Maori or Native American Indian history. These important issues will be discussed in more detail in Chapter 7.

The validity of history and its conversion into heritage is often questioned unless there is written documentation of its significance. This is problematic in itself, as clearly the working classes, ethnic minorities, women and other marginalised groups in society have traditionally not had access to the form of education that would be required in order to produce written documentation of their experiences. There may be fragments and remnants of physical evidence, or possibly oral histories, but the depiction of the past and the representation of its citizens has traditionally been dominated by educated, Western, white males. Hence, the distortion of historical accounts has sadly been inevitable for most of our documented past. Historians such as Trevor-Roper deemed much of the world's history to be 'irrelevant' and unworthy of serious study. Recent developments in history have perhaps been even more dangerous as historians have tried to omit or distort historical evidence in an attempt to construct an alternative narrative. For example, the revisionist historian David Irving tried to convince the public in the late 1990s of 'die Auschwitzluge' (the lie about Auschwitz). Luckily, his ideas were publicly and definitively disproved.

It could be argued that the growth of international tourism has led to a revival of interest in the history and heritage of native communities and remote cultures, and although tourism is sometimes referred to as a new form of imperialism, as discussed in Chapter 3, it cannot easily compete with the narrow-minded bigotry of European hegemony!

## Ethnographic museums

The development of ethnographic museums and galleries is well documented in museum studies literature. Durrans (1988) explains how ethnographic museums originally emerged as adjuncts of European expansion and colonialism; consequently representation of ethnic groups was often overtly or covertly racist or patronising. He describes how Western museums and art galleries tend to display ethnic objects in isolation, disconnected from any social context. This is in accordance with the aesthetic 'auratic' conventions of the west, whereas ethnographic museums focus more on an object's original meaning, function and purpose. Kirschenblatt-Gimblett (1998: 18) discusses in some detail the problems of displaying ethnographic collections, stating that 'Like the ruin, the ethnographic fragment is informed by the poetics of detachment'. Some interpretation may therefore be needed, as the cultural meanings attached to objects, events or traditions cannot usually be predicted from the objects alone. However, such interpretation needs to be managed carefully and sensitively, involving the appropriate ethnic community in its development.

Another problem of ethnographic museums has been the often illegal or forceful acquisition of cultural property. Although ethnographic museums are increasingly taking an ethical stance, often refusing to acquire material that has been smuggled out of the country of origin, repatriation of ethnic objects is still a sensitive and controversial issue. Simpson (1996) notes that this is particularly significant when ethnic and indigenous groups whose culture is still thriving today seek the return of their cultural property. Such objects are often symbols of their cultural identity, their future survival and cultural continuity. However, the fossilisation of ethnic and indigenous cultures is still not unusual. Cummins (1996: 92) describes how 'in many cases, museums still continued to portray African art and material culture as if it were purely ethnographic relics of the past, rather than evidence of a sentient people and a living culture'.

## Industrial heritage tourism

The postmodern pluralisation of history and its interpretation and representation have led to a growing appreciation of industrial, agricultural and popular folk heritage. In the past, people were invited to pay their respects largely to objects which had been deemed 'worthy' by a body remote from the public's own experience of life. However, the fundamental principles of postmodernism reject what are seen as evaluative prejudices, and it should be remembered that the aesthetic appeal of cultural attractions remains largely subjective. Edwards and Llurdes (1996: 343) refer to the 'aesthetics of deindustrialisation' in their discussion of industrial heritage sites in Britain and Spain, and quote Hoyau (1988: 29–30): 'Once the notion of "heritage" has been cut free from its attachment to beauty, anything can be part of it, from miners' cottages or public washouses [*sic*] to the halls of Versailles, so long as it is historical evidence.' The idea here is that 'historicity' or historical value takes priority over the aesthetics of the site. However, industrial heritage tourism also serves an important regenerative purpose.

The development of industrial heritage tourism in Britain has flourished in recent years, especially in the Ironbridge area near Telford in Shropshire. The rapid de-industrialisation of Britain in the late 1970s and 1980s created an innate sense of loss, especially among northern communities. Although Hewison (1987) criticised the British people's nostalgia for the industrial past, it is clear that there was a need to fill the gap that was left in many people's lives and livelihoods as a result of de-industrialisation. The development of industrial heritage tourism has often helped to regenerate areas in decline, boosting the local economy and contributing to people's sense of identity and self-worth. The same has been true to a certain extent of many agricultural communities across Europe.

The Northern Ireland Tourist Board is currently promoting the industrial heritage of Northern Ireland, with guided tours around old linen mills, corn mills, railways, factories, breweries, potteries, glass-cutting and -blowing workshops and so on. Similarly, the Scottish Tourist Board drew up an Industrial Heritage Strategic Plan for Scottish Tourism in 1994 to help manage and market the eighty-two sites and museums dedicated exclusively to Scottish Industrial Heritage themes, and a further sixty-two sites and museums containing an industrial heritage element as part of a much wider display of themes. Many former coal-mines in Wales are being developed for tourism, some with former miners acting as guides. Steam railways are also a popular attraction.

Although Britain has led the way in its development of industrial heritage tourism, it is clear that other European countries and communities have followed suit. For example, the French have expanded the industrial tourism market, focusing in particular on factory visits; for example, to the EDF dams and nuclear power-stations, Kronenbourg breweries, Roquefort cheese cellars, a Benedictine liqueur distillery and the Aerospatiale plant in Toulouse. In the Schio region of northern Italy, the Bureau of Culture and the Teachers' Centre of Democratic Initiatives (CICI) have started a common action of conservation and revaluation of the machinery civilisation of the Schio area, including the setting up of an outdoor Museum of Industrial Heritage. The museum includes several formative itineraries, such as wool factories, canals, sawmills, hydro-electric power-plants, spinning mills, mining and so on. A choice of six or seven thematic trails enables visitors to gain an understanding and appreciation of the revolution which gave birth to industrial civilisation.

A German tourism initiative in Nordrheinwestfalen consists of an 'Industrial Heritage Trail' through the Ruhrgebiet. The trail is centred around the 'Zollverein XII' colliery in Essen, with a network of seventeen attractions; three visitor centres; five museums; five viewpoints; linked to a calendar of cultural events in the region. An impressive Industrial Heritage Virtual Museum is also available on the Worldwide Web.

The former GDR offered a kind of 'living museum' in terms of technical installations after 1989; however, much of it was unfortunately destroyed before it could be preserved. Fascinating evidence of typically German industries still survive, such as the sugar factories around Magdeburg; potassium mines in Bleicherode; brown coal-mines around Bitterfeld; a film factory in Wolfen; and a textile factory in Cimmitschau. EU-funded initiatives have been developed around these attractions in recent years.

The Spanish have perhaps been slower to develop their industrial heritage, due largely to a narrower concept of what constitutes 'worthwhile' heritage, although Edwards and Llurdes (1996) cite interesting examples of mining tourism in Cardona and Riontinto.

## The ecomuseum concept

It is interesting to note the development of the ecomuseum concept in agricultural and rural areas of Europe. Although the concept originated in France, it has become increasingly popular in Northern Europe, especially in Scandinavia. Tunbridge and Ashworth (1996: 37) describe an 'ecomuseum' as an entire region, which 'presents its distinctive characteristics (physical as well as cultural) as a unique and indivisible heritage synthesis in which the daily lives of the inhabitants, past and present, constitute an essential integral component.' Walsh (1992: 162) states that: 'The ecomuseum is concerned to integrate all of the disciplines which are normally involved in museology, including archaeology, social history, understanding of people and places.'

The word 'ecomuseum' was coined in the 1960s in France, in the form of 'écomusée'; hence it is generally assumed that the ecomuseum was a French invention. The ecomuseum was part of a broader movement known as the 'new museology' which compelled and encouraged people to look at a certain kind of historical evidence in a different way. The approach was a radical departure from tradition, particularly because it focused predominantly on the characteristics of specific local communities, usually in rural areas. It is clear that the ecomuseum concept has participated in reshaping the definition of 'heritage' in French society (Poulot, 1994).

Hugues de Varine and Georges Henri Rivière used the site in and around Le Creusot to develop the ecomuseum concept in 1974. From the late eighteenth until the mid-twentieth century Le Creusot had been one of the most important industrial regions in France. Its prosperity had been built around the production of armaments and railway locomotives. The idea behind the initiative was that local people would not only help to create the museum, but would themselves be living exhibits in it. This Mark Two ecomuseum was to be concerned with an area of about 500 $km^2$, half industrial and half rural, divided into twenty-five communes, with a total of 150,000 inhabitants. It contained two urban communities, Le Creusot, whose income had been based on manufacturing, and Montceau-les-Mines, a coal-mining town (Hudson, 1996).

Hudson (1996) makes an interesting comparison between The Museum of Man and Industry at Le Creusot with the museum complex at Ironbridge Gorge in central England. Both are scattered sites, but with a central interpretative museum and a cluster of smaller subsidiary museums. They also have an integrated management structure. However, Ironbridge belongs to a Trust, so has always needed to raise the greater part of its funds; therefore it was designed mainly to attract visitors from outside the area. Le Creusot was brought into being to enrich the lives of the local inhabitants; therefore the income from tourism is less important. It would be interesting to end this section with a case study of the Ironbridge Gorge museum, as it was one of the first and most influential industrial heritage sites in Europe (Box 5.3). It also combines elements of both industrial and rural heritage, and embraces the ecomuseum concept.

## Box 5.3 Ironbridge Gorge Museum

Ironbridge is based around Coalbrookdale which is one of the most remarkable early industrial settlements from Britain's Industrial Revolution. It is also the area where the world's first iron bridge was built.

The Ironbridge Gorge Museum is a collective title for an area which encompasses a range of museums and sites of interest:

- The Ironbridge (the first of its kind; opened to traffic and visitors in 1781).
- The Coalbrookdale Furnace and the Museum of Iron (the furnace used coke in iron-smelting which revolutionised technology, and was crucial to the construction of the Ironbridge; the museum includes exhibits of iron objects made in Coalbrookdale, as well as a history of iron-making).
- The Jackfield Tile Museum (based on the Craven Dunnill tile works founded in 1791. The area was well known for its tile-making).
- The Coalport China Museum (based on china and porcelain manufacturing in Coalport in the 1770s).
- Blists Hill Open Air Museum (a 'living history' museum opened in 1973).

continued

- The Museum of the River (the most recent museum opened in 1989. It shows where the greatest concentrations of industrial activity were and how transport networks connected them).

Blists Hill is a good example of a site of 'living history' as was discussed earlier in this chapter. It covers fifty acres, and was originally the site of several coal-mines established in the late eighteenth century. There is a high street of shops where staff wear Victorian dress and demonstrate typical activities of the time.

West (1988) is fairly critical of the site, arguing, first, that the representation of the so-called 'community' of Blists Hill is far from being authentic. Second, he argues that the educative function is almost entirely subordinated to the opportunism of commodity exchange: 'what we have here is a complete breakdown of the distinction between history-making and money-making' (p. 57).

This is perhaps a cynical attitude to the Trust's obvious need to generate its own income. It should also be remembered that the Ironbridge Gorge Museum is generally perceived to be one of the most innovative and interesting industrial heritage sites in Europe, and is incredibly popular with both domestic and international visitors. It was the winner of the first Museum of the Year Award in 1977, followed by the first European Museum of the Year Award. However, West's comments are typical of the debates that are raging currently within the heritage field about the perceived commodification of history; the problems of authentic representation of the past; the tensions that exist between the education and entertainment function of heritage sites; and the role of tourism as a means of generating income for heritage attractions.

Source: Adapted from http://www.ironbridge.org.uk/ and West (1988)

## Conclusion

This chapter has attempted to provide an overview of some of the fascinating and complex issues surrounding the development of the heritage and museum sectors, and the role cultural tourism plays within this development. It is clear that the postmodern recognition of the need for a departure from traditional, elitist and exclusive interpretations of history has led to a number of significant changes in the way in which the past is represented through heritage. Although it could be argued that the development of the cultural tourism industry has exacerbated the commodification of heritage, and the 'heritagisation' of the past, it has also led to an increase of interest in the histories and heritage of regional, ethnic, indigenous and other previously marginalised groups (e.g. women and the working classes). The future success of the sector is arguably dependent on a more inclusive, democratic and multicultural approach to both the interpretation and representation of history. Only then will society recognise the diversity and multiplicity of cultural groups that have contributed not only to its past, but to its present and future.

# 6 The globalisation of heritage tourism

> Heritage sites and buildings are not just important because of what they reveal about the past. . . . Nor are they just fine parts of a human-created landscape that are pleasing to the eye and interesting to the intellect. They are examples that we carry with us into the future. We can learn from them, we can teach from them, we can inform our future choices by understanding them. In a very real sense, heritage is as much about the future as it is about the past.
>
> (Smith, 1998: 69)

## Introduction

The aim of this chapter is to analyse the relationship between tourism and heritage within a global context. Emphasis will be placed in particular on the 'universalisation' of heritage through the development of the World Heritage Site list, and the growth of global tourism development. Whereas Chapter 5 considered many of the political ramifications of cultural tourism development, focusing in particular on interpretation and representation, this chapter will consider some of the conservation and management issues that are pertinent to the heritage tourism sector. The development of global blueprints for the management of heritage sites can be problematic and sometimes dissonant to local communities, yet they help to ensure the future preservation and conservation of heritage sites of universal value. Some of the tensions implicit in this process will be discussed in more detail, as will some of the spatial issues, which have helped to shape the geographical development of heritage tourism.

## The globalisation of heritage

Whereas the previous chapter emphasised clearly the relationship between history and heritage, this chapter will consider the geography of heritage, including an analysis of the increasing globalisation of the world's heritage attractions, largely through tourism. As discussed elsewhere in the book, debates about globalisation and postmodernism have often focused on the issue of 'time–space compression'. As stated by Harvey, postmodernism focuses on both temporal and spatial issues and is therefore 'a historical-geographical condition of a certain sort' (1990: 328). The concept of time–space compression is applied frequently to the tourism industry, and is increasingly relevant to the heritage industry, especially as it becomes more 'mediatised' on a global scale. So, although we can refer to a geography of heritage (Graham *et al.*, 2000), spatial and temporal considerations are inextricably linked.

In temporal terms, many authors have emphasised the fact that we must think not only in terms of the past, but also the present and future when planning and managing heritage.

For example, Herbert (1995a: 215) states that 'History must not be compartmentalized as something which belongs in the past, it is part of a continuity and is intrinsic to our modern heritage', and Gruffudd (1995: 50) suggests that 'Historical narratives reveal contemporary anxieties, and contemporary desires are fulfilled in the presentation of the past'. Clearly the actions of the present help to shape the preservation, conservation and management of heritage in the future. As discussed in previous chapters, selective, exclusive and biased approaches to the interpretation of history have often resulted in a limited representation of the past, which fails to take account of a multiplicity of perspectives. This can be partially rectified in the present and future.

However, there has been some concern that the protection of heritage has somehow been responsible for destroying the present. Urry (2002: 99) states that:

> The protection of the past conceals the destruction of the present. There is an absolute distinction between authentic history (continuing and therefore dangerous) and heritage (past, dead and safe). The latter, in short, conceals social and spatial inequalities, masks a shallow commercialism and consumerism, and may in part at least destroy elements of the building or artefacts supposedly being consumed.

He cites Hewison, who declared that history needed to be protected from the conservationists, who are sure to turn it into heritage or 'bogus history' (1987: 144). Larkham (1995: 86) differentiated between the concepts of preservation, conservation and exploitation as follows:

- *Preservation* involves the retention, in largely unchanged form, of sites and objects of major cultural significance.
- *Conservation* encompasses the idea that some form of restoration should be undertaken to bring old buildings and sites into suitable modern use.
- *Exploitation* recognises the value of heritage sites, particularly for tourism and recreation, and encompasses the development of existing and new sites.

He notes the criticisms that have met trends like 'facadism', which involves the development of new structures and modern designs behind original facades. This can lead to the loss of the historic totality of buildings and misrepresentation of townscapes. Of course, in the case of cities such as Warsaw, whose main squares were almost wholly destroyed during a war, facadism may afford the only option in terms of restoration of the original features.

Another much criticised form of architecture, which is perhaps more likely to be linked to urban enhancement or regeneration schemes than conservation, is 'pastiche'. This phenomenon is often believed to symbolise the postmodern era of eclecticism, where 'the geography of differentiated tastes and cultures is turned into a pot-pourri of internationalism' (Harvey, 1990: 87). Jameson (1984) lamented the depthless and contrived nature of postmodern architecture, which made extensive use of a hotchpotch of different styles. Harvey (1990) describes how a veil is effectively drawn over real geography with the creation of artificial places (e.g. China towns, Arab Quarters, Little Italies), and the staging of heritage events and spectacles. As discussed in Chapter 9, the transformation of space, especially spaces of historic significance, has prompted a furore over the apparent 'heritagisation' of space whereby places are 'de-historicised' (Walsh, 1992). However, Harvey (1990) suggests that many spaces have been converted to reflect the 'transplanted' cultures of their inhabitants (e.g. Italian immigrants or other diasporic groups). Although this is again facadism rather than realism, it becomes like a symbol of a displaced or relocated heritage.

In spatial terms, heritage may be defined in relation to a number of geographically bounded areas such as neighbourhoods, destinations, regions or nations. Alternatively, it may be defined socially or culturally in relation to real or imagined communities and what is significant to them:

> History, commemoration and conservation are all . . . implicitly political. What a self-defined group or a nation seeks to preserve, and to represent to others, allows us to understand something of what a particular imagined community thinks it is.
> (Gruffudd, 1995: 49)

The links to personal or collective identity construction here are evident. Hewison (1987) describes how the preservation of the past is linked closely to the preservation of the self and a sense of identity. However, we should not lose sight of the fact that we are not only defined according to our past, but also to our contemporary lives and our future aspirations. Evans (2001: 13) notes that 'the extent to which cultural heritage should be prioritised over contemporary culture and living art is a complex and ultimately political issue'. In cultures where past, present and future are not so readily compartmentalised, such a distinction is less relevant. However, the Western tendency to fossilise culture and to wallow in nostalgia means that the preservation or conservation of heritage has often been paid disproportionate attention.

The complexity of local, regional and national identity construction has been discussed elsewhere in this book. Collective constructs of identity might include class, gender, ethnicity or nationalism, all of which are fundamental to our individual and collective perceptions of our heritage. Graham *et al.* (2000: 90) suggest that 'the relationship between heritage, identity and place can now be seen as intensely heterogeneous and full of nuances and ambiguities'. Belonging to a place is clearly fundamental to identity construction, but displacement has been common in the postmodern era, not just in terms of travel and tourism, but also immigration or exile, for example. Hence identities are in constant flux, or become hybridised. There is a kind of multilayering of place and identity consisting of potentially conflicting supranational, national, regional and local expressions.

Graham *et al.* (2000: 238) describe how (cultural) tourism development has been instrumental in 'globalising' heritage:

> It can be argued that the inexorable growth of foreign tourism, and the importance of culture, heritage and art to that industry, is the most powerful expression of the existence of a common global heritage as the property of all peoples. Every international tourist is asserting the existence of a world heritage and the right of a global accessibility to it, as well as more mundanely selecting the content of that heritage and continuing its support.

Urry provides a detailed discussion of the globalisation of the gaze, and he outlines the many gazes that have been adopted by tourists throughout the history of tourism and in recent times: 'There has been a massive shift from a more or less single tourist gaze in the 19th Century to the proliferation of countless discourses, forms and embodiments of tourist gazes now' (2002: 160). He describes how heritage has traditionally been more central to the gaze in Britain than in other countries; however, it is clear that heritage tourism is a global growth sector like most other forms of cultural tourism.

Evans (2001: 226) emphasises the need to reconcile national and local needs as part of global heritage tourism management:

> The focus on world and symbolic heritage sites in the cities of both developed and developing countries requires that a balance be struck between local and

national imperatives – qualities of life, economic and physical access, minimising gentrification effects and the imposition of 'staged authenticity' in terms of the heritage that is conserved.

The challenge of achieving such a balance will be discussed in more detail in the latter part of this chapter, which focuses on some of the management issues that are central to the development of heritage tourism.

## An overview of cultural heritage

Although heritage may be conceptualised as meaning rather than artefact (Graham *et al.*, 2000), the remainder of this chapter focuses predominantly on material rather than intangible heritage, aspects of which are discussed in other chapters. While it is recognised that heritage can be defined as being both 'natural' and 'cultural', especially in terms of the World Heritage Site list, this chapter focuses mainly on cultural heritage in accordance with the book's central theme. Although such categorisations are by no means clear-cut (see, for example, cultural landscapes), there is not scope to discuss the concept of heritage in its broadest sense here. However, many of the more general theories of heritage conservation and heritage management are just as applicable to natural as to cultural heritage sites.

Feilden and Jokilehto (1998: 11) define cultural heritage quite broadly as 'containing all the signs that document the activities and achievements of human beings over time'. This is a useful definition, and one that is used for UNESCO's World Heritage List. It recognises cultural heritage as a broad concept relating to the development of contemporary society, its values and its needs. Cultural heritage does not relate only to tangible works of art or architecture as part of the built environment, but also to the more intangible aspects of people's lives, traditions and customs. ICOMOS (1999) defined heritage as:

> a broad concept and includes the natural as well as the cultural environment. It encompasses landscapes, historic places, sites and built environments, as well as biodiversity, collections, past and continuing cultural practices, knowledge and living experiences. It records and expresses the long processes of historic development, forming the essence of diverse national, regional, indigenous and local identities and is an integral part of modern life. It is a dynamic reference point and positive instrument for growth and change. The particular heritage and collective memory of each locality or community is irreplaceable and an important foundation for development both now and in the future.

This is an interesting quotation, as it demonstrates the diversity of sites which are described as heritage. It also recognises the significance of indigenous and local identities, especially in an increasingly globalised world. The protection, conservation, interpretation and presentation of heritage are important challenges for both present and future generations. The political dimension of heritage and its connection to a sense of place and identity are central to debates about cultural development. Heritage plays an important political, social and educational role. The conservation, protection and management of heritage are therefore being recognised as increasingly important to the future of society.

## The relationship between heritage and cultural tourism

The relationship between tourism development and heritage management is a complex and sensitive one, but it is by no means a recent phenomenon. As stated by Graham *et al.* (2000: 238):

> This close symbiosis between heritage and tourism is not new. The desire to conserve and the desire to visit have always mutually stimulated the history of the heritage conservation movement in an inextricable relationship of cause and effect.

The relationship between tourism and heritage is often perceived as being fraught with problems, rather than harmonious and symbiotic. Although Ashworth (1995: 71) questions 'the naïve assumptions of harmony' that exist among those who argue that tourism and heritage necessarily enjoy a symbiotic relationship, he recognises the mutual benefits of heritage tourism development:

> Tourism makes use of the conserved artefacts of the past, thereby acquiring a patina of artistic patronage and educational worth, while the conservation lobby acquires justification and political support, as well as the possibility of a much-needed financial contribution.
>
> (Ibid.)

Again, there are a large number of books and articles devoted to the relationship between tourism and heritage; therefore this book does not set out to duplicate these. However, it does aim to provide a useful synthesis of some of the key issues, such as those listed below:

- 'Ownership' of heritage.
- Questions surrounding appropriate use of heritage.
- Access versus conservation.
- Heritage as an industry, business or product.
- Heritage as entertainment.
- Heritage as formal or informal education.
- The interpretation of heritage.
- Heritage and authenticity.
- Heritage and representation.

Many of these issues have been discussed in more detail in Chapter 5 and elsewhere in this book. This chapter is concerned particularly with use and ownership of heritage, and some of the debates surrounding the conservation and management of sites. The following list suggests examples of the types of heritage sites that have become cultural tourism attractions in recent years:

- *Built heritage attractions* (e.g. historic townscapes, architecture, archaeological sites, monuments, historic buildings).
- *Natural heritage attractions* (e.g. national parks, cultural landscapes, coastlines, caves, geological features).
- *Cultural heritage attractions* (e.g. arts, crafts, festivals, traditional events, folk history museums).
- *Industrial heritage attractions* (e.g. mines, factories, mills).
- *Religious sites and attractions* (e.g. cathedrals, abbeys, mosques, shrines, pilgrimage routes and cities).

- *Military heritage attractions* (e.g. castles, battlefields, concentration camps, military museums).
- *Literary or artistic heritage attractions* (e.g. houses, gardens or landscapes associated with artists and writers).

This is not a definitive list, but it does give a good indication of the kinds of heritage attractions that are becoming popular with cultural tourists. Clearly, the conservation and management issues that are relevant to each site will vary considerably, as will the profile of visitors. However, generalisations can perhaps be made about the types of visitors who are usually described as being 'heritage tourists':

- Better than average education.
- Age groups 20 to 30 or 45 to 60.
- In the older age category, or above-average income.
- An actual or aspirant member of the middle class.
- Travelling without children.
- Experienced in foreign travel.
- Chooses catered accommodation forms.
- Relatively high per diem expenditure.

(Adapted from Tunbridge and Ashworth, 1996)

Heritage tourists are often believed to be a 'better class' of tourist because they tend to spend more money in the local economy of a destination and they are supposedly sensitive to the local culture, customs and traditions of the host community. They are also more likely to have some awareness of environmental and conservation issues.

However, tourist numbers at some of the world's heritage attractions, especially World Heritage sites and historic cities, are becoming a cause for concern. Mass tourism is no longer a phenomenon that occurs only in beach resorts. Many major attractions are finding it increasingly difficult to balance the conservation of the site, maximising access and optimising the visitor experience. The following list gives some good examples of sites that are suffering as a result of tourism development:

- Cultural and historic cities (e.g. Venice, Kracow, Prague, Oxford).
- National Parks (e.g. Lake District, Yellowstone Park).
- Archaeological sites (e.g. Ephesus, Pompeii, Hampi).
- Individual sites and monuments (especially World Heritage Sites, such as the Pyramids, Taj Mahal, Stonehenge, Canterbury Cathedral).
- Intangible heritage (e.g. native customs, traditions, rituals).

As a result of the perceived fragility of the world's heritage, a number of global measures have been taken to protect such sites and attractions from the ravages of 'mass cultural tourism'. The most significant of these is the establishment of the World Heritage List. The following section will provide an overview of the designation, status and management of World Heritage sites. This offers a useful framework for an analysis of key issues in global cultural heritage, and of the dynamic interaction between tourism and cultural heritage. Although it is recognised that each World Heritage Site is unique in nature, management is governed by a set of international guidelines that are applied in a local context.

## Organisational framework for World Heritage Sites

UNESCO has an overarching role in the protection and management of World Heritage Sites; however, ICOMOS (the International Council on Monuments and Sites) is the body responsible for the management of cultural heritage whose primary objective is the conservation/preservation of monuments, groups of buildings and sites. Through its membership and the exchange of information and expertise, ICOMOS forms an international network that defines, improves, promotes conservation/preservation principles, standards, research, responsible practice and innovation.

In 1972, UNESCO adopted the World Heritage Convention in order to protect cultural and natural heritage worldwide, and to provide organised international support for the protection of World Heritage Sites. The member states, which contribute funding to UNESCO, ratify this document, which is then managed by the World Heritage Committee. The Convention protects heritage sites of 'outstanding universal value'.

The newest category on the World Heritage List is that of mixed sites or cultural landscapes, which were admitted to the list in 1996. In some cases, sites meet both natural and cultural criteria; therefore it is more appropriate to give them dual status. UNESCO defines several types of cultural landscape. These are:

1. designed and created intentionally by man
2. organically evolved
3. a relic (or fossil) landscape
4. a continuing landscape
5. an 'associative' landscape.

The final category is often linked with aspects of intangible heritage of a religious or spiritual nature (e.g. aboriginal landscapes). It may also be landscapes associated with famous people, such as authors, artists or poets. Overall, the cultural landscape category considers the whole environment, rather than individual sites, including the natural and built environment, and any resident communities.

In addition to the designation of cultural sites, natural sites and cultural landscapes, the World Heritage Committee also decides when sites should be inscribed on the World Heritage in Danger List (Box 6.1). These threats may be natural (e.g. earthquake, flood, volcano) or man-made (e.g. civil war, deforestation, tourism). Such sites are monitored and afforded special protection for such time as they are still considered to be under threat. Sites are not usually removed from the World Heritage List unless the site has deteriorated to the extent that it has lost the characteristics which determined its inclusion on the List in the first place.

## World Heritage Site designation

The procedure for World Heritage Site designation is complex and lengthy. The process of nomination, evaluation and inscription takes at least eighteen months. Governments must submit a dossier for a particular site. Sites are then required to produce detailed historical and archaeological records, as well as submitting a management plan. A committee considers the information and decides whether the site should be nominated. Governments must have the required legal and regulatory frameworks and management mechanisms to support nomination.

All potential World Heritage sites must also meet tests of authenticity in design, material, workmanship or setting, and, in the case of cultural landscapes, their distinctive character

## Box 6.1 World heritage in danger: the Taj Mahal

Agra is a small town in the Indian state of Uttar Pradesh with little to distinguish it from any other northern Indian city, except that it is home to three of India's and the world's best-known and most impressive World Heritage Sites. These are: The Taj Mahal, Agra Fort and the deserted city of Fatehpur Sikri.

These sites have become threatened increasingly by urban development and tourism. In the early 1980s, the World Health Organisation classified Agra as a 'pollution intensive zone' with its coke-based industries and vehicle emissions. There was great concern that the sulphur dioxide in the air had been eroding the buildings, especially the Taj Mahal, which is becoming discoloured and the marble is flaking. In addition, around ten million people view the Taj each year, so the pressures of tourism are also taking their toll.

As a consequence, in 1993, the Supreme Court ordered that new industries should be banned within a 10,000 sq ft exclusion zone, and that existing coke-based industries should move out or switch over to gas. The Court also ordered that a proposed bypass highway be built to reduce the level of vehicle emissions in the centre of Agra, and vehicles have now been prevented from entering the precincts of the Taj, and parking is restricted. In addition, the Taj is now closed on Mondays for cleaning.

However, it is estimated that air pollution is still five times higher than the Taj can withstand without damage, and the monument is unlikely to decline in popularity in the near future. The

**Plate 6.1** The Taj Mahal: global heritage in danger

Source: Author

only deterrent is likely to be the relatively exorbitant entrance fee for foreign visitors, which is now $20.

A further threat that has been identified are the plans to surround the Taj Mahal with cyber cafés, fast-food restaurants and shopping malls. The development is part of a deal signed by the government with the Taj Group in July 2001, India's largest hotel chain, and is supposedly going to help renovate and upgrade the site. This will include various 'beautification' projects along the river, and a sound and light show. Although spokesmen for the Taj Group have responded to criticism by saying that their only motives are altruistic in their desire to preserve India's heritage, traditionalists may well prove to be justified in their suspicion and outrage.

Sources: Adapted from Baig (2001); Berry (2001); Taj Mahal (2001)

and components (Shackley, 1998). This process is clearly complicated when local communities are part of the landscape, and there are inevitable conflicts between conservation, tourism and local needs.

The quest for World Heritage Site status appears to be growing, as both developed and developing nations compete for the acquisition of this much-coveted global accolade. This is partly a result of a strategic shift in World Heritage emphasis over the past decade towards a less Eurocentric and more inclusive and democratic approach to designation. In 1994, the World Heritage Committee called for:

> rectification of the imbalances on the List between regions of the world, types of monument, and periods, and at the same time a move away from a purely architectural view of the cultural heritage of humanity towards one which was much more anthropological, multi-functional, and universal.
>
> (World Heritage Committee, 1994: 4)

Some of the more developed countries were asked to slow down their rate of nomination so as to address the perceived imbalance and the apparent favouring of material over non-material cultures (Pocock, 1997a). The addition of the category of cultural landscape to the World Heritage List in 1996 has helped to take better account of non-material and indigenous cultures. Equally, more industrial landscapes have started to be included on the World Heritage List (Box 6.2) (see e.g. Smith, 2000). Graham *et al.* (2000) suggest that critics who have opposed these trends are largely Eurocentric, and that the concept of equity of distribution of World Heritage status does not necessarily override the site's intrinsic merits. However, they are cynical about the apparently unlimited expansion of the List, which they see as threatening its credibility.

The 'globalisation' of heritage could be perceived as being a source of dissonance (e.g. Tunbridge and Ashworth, 1996; Graham *et al.*, 2000), especially in terms of local ownership (collective or otherwise) and interpretation. Politically, World Heritage status necessarily transcends national boundaries. As stated by Graham *et al.* (2000: 97):

> Internationally, UNESCO, particularly through the medium of World Heritage Sites, pits the ideas of universal heritage values (themselves at times the subject of intractable dissonances) against the self-interest of the various host states, chiefly concerned with national-scale priorities.

Tunbridge and Ashworth (1996) have argued that all heritage is dissonant, and that the designation of sites is often used by nation states as a means of engaging in national aggrandisement, as well as being an assertion of Eurocentricity on the part of international

## Box 6.2 Blaenavon industrial landscape

Blaenavon is an industrial World Heritage Site, which is located in South Wales. It was one of the world's largest producers of iron and coal in the nineteenth century, and is seen as making an extremely important contribution to the Industrial Revolution. Evidence of the area's past may be seen *in situ*, including coal and iron ore mines, quarries, a primitive railway system, furnaces, the homes of the workers, and the social infrastructures of their community.

The property was inscribed on the World Heritage List in 2000 according to the following criteria:

- The Blaenavon landscape constitutes an exceptional illustration in material form of the social and economic structure of the nineteenth-century industry.
- The components of the Blaenavon industrial landscape together make up an outstanding and remarkably complete example of a nineteenth-century industrial landscape.

The main reasons for inscribing this industrial landscape are to raise awareness of the social and economic significance of this area and its communities, and to assist with regeneration. Like most industrial heritage sites, Blaenavon has suffered from considerable economic decline in recent years; therefore the preservation of the site serves as a physical reminder linking the present and the past. Many unemployed miners are becoming more involved in tourism development in order to interpret and represent their collective heritage. Although it could be argued that World Heritage Site status affords local communities few benefits, it is worth noting that they are far more likely to feel an affinity with an industrial or agricultural landscape than with a castle or palace. Hence local pride in past traditions can be revived and strengthened through conservation and tourism development.

Blanaevon has sometimes been referred to as 'The Taj Mahal of the Valleys', and although cynics would no doubt argue that the site cannot hope to compete with some of the more aesthetic World Heritage sites, it is in cases like these that 'historicity' or historical value are the main reasons for inscription. This comes back to the increasingly democratic approach being adopted for World Heritage Site designation. Rather than being a mere beauty pageant, the initiative serves to recognise sites of universal value in a broad range of contexts.

Source: Blaenavon Industrial Landscape (2001)

agencies. They emphasise the discordance between heritage identity and local developments. Clearly, the development from a local monument, park or town to an internationally recognised symbol can be problematic. The subtle social, cultural and political implications of the transformation of sites into global tourist attractions, and, by implication, visual signifiers or branding tools, requires further analysis, especially in terms of local dissonance.

The following section will address some of the broader issues relating to the significance of World Heritage Site status, questioning the diversity of motives that appear to underpin the quest for designation.

## The significance of World Heritage Site status

The significance of World Heritage Site status for the sites themselves, national and local government, relevant authorities, local communities, and visitors to the site may differ enormously. Any analysis of the implications and significance of World Heritage status is rendered problematic by the diverse nature of the sites, and is complicated further by the range of contexts in which they are located. Although all sites may share common management dilemmas relating to conservation, access and visitation, their social and political significance is highly dependent on the type of site and its location within a specific context. For example, the social and political significance of a designated indigenous cultural landscape in a developing country is unlikely to be directly comparable to that of a European historic city. In social or political terms, the local planning, management or regulatory framework varies enormously, as does the composition of the local community, which may be as heterogeneous as the sites themselves. For example, a close-knit, mono-cultural tribal group living within a national park cannot easily be compared to the cosmopolitan, multi-ethnic population of a city.

However, irrespective of the context or perspective, World Heritage Site status seems to serve a number of common purposes. Graham *et al.* (2000: 243) suggest that sites are afforded international protection and levels of expertise, both for conservation and management:

> The World Heritage List has given a profile and a momentum to the idea of global heritage where none existed thirty years ago. Global declarations and mechanisms have provided standards to encourage best practice and access to a world reserve of technical and professional expertise. Furthermore, an ongoing debate is engaging with the meaning and potential achievements of world heritage.

World Heritage status is perceived as bringing enormous prestige at global and national level, as well as impacting upon future planning decisions at local level. Although the status could be perceived as being mainly honorific, the practical benefits are increasingly being recognised and exploited. In all cases, the preservation of these universally unique sites is of primary importance; therefore emphasis is placed on World Heritage status as a conservation tool. As stated by Lowenthal (1998: 228): 'Global accolades deter despoliation and lend sanctuary to rare legacies.' World Heritage Sites, especially composite sites (such as historic towns), occupy a clearly defined space within a boundary and buffer zone, which afford additional protection, and curtail inappropriate local developments which might encroach on or compromise the aesthetic or historic value of the site.

It is sometimes assumed that World Heritage Site designation will lead to increased publicity and higher visitor numbers (Feilden and Jokilehto, 1998). However, Shackley (1998) suggests that a significant increase in visitors is not inevitable, but is dependent on the marketing of the site and its approach to access. The issue of access is important and complex, as stated by Borley (1995: 187): 'There is always tension between the need to protect the ambiance or environment of a cultural site and the need to provide access. The latter, whilst important and desirable for education of the public, must never be allowed to diminish the site itself.' UNESCO's (1972) policy of making World Heritage Sites available to the widest possible public is a democratic and inclusive one, but care must be taken to protect the physical structure of many of the more fragile sites which can be severely degraded by excess visitation.

Few sites are, however, in a position to forgo the economic support afforded by tourism development. This does not, of course, necessarily imply a harmonious relationship between the conservation and tourism development functions of World Heritage status, and this often fraught relationship is well documented (e.g. Boniface, 1995; Shackley, 1998). It is clear that the need to achieve the fine balance between conservation and visitation is common to all World Heritage Sites (Shackley, 1998), and there is also the added pressure of designated sites being forced to act as global models of good management practice. As stated by Boniface (1995: 106): 'All the World Heritage Sites have a special obligation to take a lead, to show themselves to be world models in the appropriate management of cultural sites for tourism.' Pocock (1997b: 383) refers to 'the burden of inscription', stating that: 'The conferring of World Heritage status . . . brings responsibilities as well as perceived benefit, for inscription brings with it the duty of protection and a management programme which is subject to monitoring.'

In terms of education, World Heritage Sites are seen as being important for what they can teach local, regional and even international communities, not only about their past, but also about their present and future. World Heritage Sites should therefore be under obligation to provide adequate educational and interpretation facilities so that the public can gain maximum benefit from their visit. Shackley (1998: 1) suggests that 'visiting a World Heritage Site should be a major intellectual experience, on a different scale from visiting some theme park'.

Finally, the symbolic aspect of World Heritage Site status is significant. Shackley (1998: 205) states that 'World Heritage Sites have the highest visibility of any cultural attractions in the world, and possess a symbolic value which may be disproportionate to their size or beauty. They are symbols of our history, cultural icons whose importance transcends their current political status.' The inscription of heritage sites on the World Heritage List elevates them to the status of global icon and national treasure, with all the political and financial support that this entails.

Overall, the benefits of World Heritage Site status can be significant. The designated site often receives some extra funding for conservation and development (albeit a nominal amount). It can also lead to an increase in the site's status locally, nationally and internationally. This may encourage tourism, depending on how well the site is promoted. Local people may feel an increased sense of pride in their environment, which can strengthen community cohesion and identity. The problem is that visitor numbers can exacerbate the site's conservation problems, especially if it is a fragile or sensitive environment. Local people can feel alienated from the site or 'disinherited', especially if it becomes more of a global tourism attraction. Questions of ownership, participation and empowerment become central to the management of sites, especially cultural landscapes, where local people are an integral part of the site. Sites clearly need to be managed carefully and sensitively in order to maximise the benefits of World Heritage status for all stakeholders.

## The management of World Heritage Sites

Each World Heritage Site is required to have a management plan, which outlines its policy towards conservation, visitors and local issues. Although UNESCO and ICOMOS provide a comprehensive set of universal guidelines, it is recognised that the management of World Heritage sites requires a diversity of approaches, as emphasised by Lord Montagu of Beaulieu in a House of Lords debate in 1993:

> With regard to the more general question of World Heritage Sites, they are of widely different kinds. Blenheim is, for example, a single historic site, whereas Bath is a complete city. Durham is a great cathedral in a city, whereas Stonehenge is a site set in an archaeological landscape. It is therefore impossible to lay down specific rules for World Heritage Sites as different forms of approach are needed for each management plan.
>
> (Wheatley, 1997: 9)

Shackley (1998: 1) also emphasises the unique nature of World Heritage Sites, but she recognises the common ground, stating that:

> No two World Heritage Sites are (by definition) alike but all share common problems such as the need for a delicate balance between visitation and conservation. All are national flag carriers, symbols in some way of national culture and character. Most are major cultural tourism attractions of their country and some (such as Stonehenge, the Pyramids, the Great Wall of China) are powerfully evocative symbols of national identity, universally recognized. World Heritage is a fragile non-renewable resource, which has to be safeguarded both to maintain its authenticity and to preserve it for future generations.

The diversity of World Heritage Sites means that their management cannot be standardised too much. Management plans must aim to address the unique features and specific nature of individual sites. Clearly, managing the balance between conservation and visitor access is a central concern. The authenticity of the site should also be preserved through appropriate and sensitive interpretation. It is interesting to note the *symbolic* significance of World Heritage sites as being associated with national identity. For indigenous communities, the site may also form part of their sense of regional or local identity.

Cunliffe (1996: 266) describes site management as 'A comprehensive planning and management process which ensures that the conservation, enhancement, and maintenance of a heritage site is deliberately and thoughtfully designed to protect its cultural significance for present and future generations.' This quotation adheres quite closely to definitions of sustainable tourism (e.g. ICOMOS), which are concerned chiefly with preserving cultural heritage for both present and future generations. Site management is also viewed as an integrated process which considers all aspects of the site, including conservation, preservation, visitation and interpretation.

Shackley (1998: xiii) emphasises the importance of visitor management for World Heritage sites, but she also acknowledges its limitations:

> Many sites are fragile, threatened not only by natural erosion, but also, increasingly, by the large numbers of visitors who wish to see them. Visitor management has become a new and as yet inexact science, which aims to balance the needs and requirements of the visitor with the potential impact that the visitor may have on fragile buildings or artefacts.

This is a useful quotation, since it outlines one of the key management tools that are being used for the management of World Heritage Sites. Many sites are becoming increasingly popular precisely because they have attained World Heritage status. One of UNESCO's main aims is to increase access to World Heritage Sites, a philosophy that could be seen to be problematic for some conservation practices. Visitor numbers and flows clearly need to be controlled, regulated and minimised where appropriate, both for conservation reasons and in order to enhance the visitor experience. Visitor behaviour also needs to be monitored, especially when sites are especially fragile or sensitive.

The ICOMOS International Cultural Tourism Charter (Box 6.3) recognises the need for heritage to be conserved and communicated to its host community and tourists. Although that heritage is managed according to internationally recognised principles, the responsibility usually falls to a particular community or custodian group. The management of tourism within the context of that heritage is a key concern. Tourism should be seen to benefit the host communities, and to provide a means and motivation for them to care for and maintain their heritage and cultural practices. The sustainable development of tourism is clearly dependent on the involvement and empowerment of local communities.

It is evident that one of the key issues here seems to be managing the balance between conservation of, and access to, cultural heritage. This is one of the most difficult

## Box 6.3 ICOMOS International Cultural Tourism Charter 1999–2002

### Objectives of the Charter:

- To facilitate and encourage those involved with heritage conservation and management to make the significance of that heritage accessible to the host community and visitors.
- To facilitate and encourage the tourism industry to promote and manage tourism in ways that respect and enhance the heritage and living cultures of host communities.
- To facilitate and encourage a dialogue between conservation interests and the tourism industry about the important and fragile nature of heritage places, collections and living cultures including the need to achieve a sustainable future for them.
- To encourage those formulating plans and policies to develop detailed, measurable goals and strategies relating to the presentation and interpretation of heritage places and cultural activities, in the context of their preservation and conservation.

Source: ICOMOS (1999)

management issues to resolve in the field of cultural heritage tourism. It is important to consider the principles of sustainable development when formulating management plans at site or landscape level. The role of the local community is pivotal, especially when considering the interpretation or representation of their cultural heritage.

The case study outlined in Box 6.4 gives examples of some of the management problems that the high profile and internationally renowned World Heritage Sites in Egypt have had to face over the years.

## Box 6.4 Managing World Heritage Sites in Egypt

Egypt's World Heritage Sites are some of the best known and best loved World Heritage Sites in the world. These include the Pyramids in Giza, the Valley of the Kings and Queens, Abu Simbel, and the temples in Luxor and Karnak. It is worth noting that the preservation of the temples at Abu Simbel was one of UNESCO's greatest achievements in terms of saving World Heritage in Danger. The creation of the Aswan Dam looked set to drown a number of significant ancient monuments, but UNESCO campaigned for funds in the 1960s and managed to rescue the monuments. They cut and rebuilt a number of temples, saving a total of fourteen, and salvaged many other treasures which were placed in museums.

However, since that time Egypt has had its fair share of problems, not least the horrific terrorist attacks on tourists by Islamic Extremists throughout the 1990s. The massacre of fifty-eight tourists at the temple of Hatshepsut in Luxor was perhaps the worst example of this. As a consequence, tourism declined dramatically, which has had a significant impact on Egypt's economy, and thus indirectly on available funds for conservation. Egypt's main sites are now guarded by tourist police and escorts who accompany tour groups, which can be simultaneously reassuring and disconcerting. Certain transport routes are restricted, including the cruise from Cairo to Aswan and back, and the only way to reach Abu Simbel is currently by aeroplane, as the road from Aswan has been closed to tourists for security reasons.

Egypt's World Heritage Sites are now managed much better than they used to be, and environmental and visitor management issues are being increasingly addressed. A short spell

**Plate 6.2** Temple of Hatshepsut, Egypt. Site of tourist massacre in the 1990s

Source: Author

on the World Heritage in Danger List forced the Egyptian government to address some of the key management issues at sites like the Pyramids, as specified by the World Heritage Committee. However, it is still quite shocking for the first-time visitor to witness the urban sprawl and high levels of pollution that are encroaching on the Pyramids in Giza (Cairo being one of the most polluted cities in the world).

Nevertheless, tourists are no longer allowed to climb the Pyramids as they once were, and entrance to the inside of the Pyramids is now restricted because the breath of too many visitors was causing excessive erosion. Visitor numbers are also restricted inside the tombs in the Valley of the Kings and Queens in Luxor, and photography is prohibited. It might be argued that visitor facilities and interpretation are not what they could be at either site, but at least conservation issues are being gradually addressed.

## Sustainable tourism development in historic towns

The management of tourism in urban environments and historic cities clearly requires a more integrated approach than individual heritage sites. Policies and planning decisions need to be integrated into a broader urban development context, and conflicts of interest between urban planners, tourism developers, conservationists and the local resident population need to be addressed. Historic towns are inhabited by thriving, working communities, and should not be viewed simply as historic attractions to be 'fossilised' or turned into 'living museums' by the tourism or heritage industries for the benefit of visitors.

Organisations such as ICOMOS, the English Historic Towns Forum, the European Association of Historic Towns and Regions, and the Walled Towns Friendship are actively involved in the development of sustainable tourism in historic towns. A recent publication produced by the English Historic Towns Forum, English Tourism Council and English Heritage sets out clearly many of the main issues that need to be addressed in historic towns. The main principles of sustainable tourism that they embrace are that:

1 It must not adversely affect the environment.
2 It must be acceptable to the community.
3 It must be profitable for business.
4 It must satisfy the visitor.

Other key issues are summarised in the following list:

- *Enhancing a sense of place*: highlighting special and unique characteristics of a town; attention to aesthetic detail (e.g. architecture, street furniture, shop fronts); instilling a sense of civic pride and 'ownership' in local communities; strengthening local identity; appropriate promotion and image creation.
- *Strengthening the evening economy*: late-night shopping; special early evening promotions in bars, restaurants and cafés; street markets and entertainment; festivals and events.
- *Attention to green spaces*: creation or improvements of parks, gardens and picnic areas; elimination of pollution and litter; maintenance of footpaths and walkways; floral enhancements, such as borders and hanging baskets.
- *Safety and security*: installation of CCTV; good street lighting; police presence; busy streets and a thriving evening economy.
- *Transport planning*: easy access; pedestrianisation; car-parking provision; coach dropping-off points; park and ride systems; minimisation of congestion and pollution; good road signage; creation of cycle lanes and bicycle hire; water transport.
- *Visitor management*: good signage; interpretation; orientation (e.g. TICs, 'Meeters and Greeters', Welcome Host schemes); guide services; visitor services and facilities (e.g. toilets, litter-bins, public phone boxes); disabled provision; town centre trails.
- *Marketing and promotion*: image creation; targeting appropriate markets; provision of information; developing special packages; addressing seasonality; de-marketing where appropriate.
- *Co-ordination*: involvement of public, private, voluntary sectors and local residents; local and regional tourism and cultural consortia; steering groups; environmental and transport agencies; local businesses; marketing bodies.

(Adapted from EHTF, 1999)

It is often difficult to resolve many of the conflicts that arise in the context of historic town management. For example, the pedestrianisation of a town can serve the purpose of minimising congestion and pollution, helping to prevent the erosion of historic buildings, and enhancing the experience of local residents and visitor alike, However, the rerouting of traffic can often create bottlenecks outside the town centre. Access is reduced for taxis and buses, and residential areas often suffer from increased congestion and pollution. It can also create problems for local traders and small businesses in the town centre that traditionally rely on deliveries to their door.

In terms of marketing, many historic towns have sometimes had to engage in *de-marketing* because their centres have become overly congested and the historic fabric is threatened. Many towns are keen to attract smaller numbers of high-spending, educated cultural or business tourists, rather than low-spending backpackers and language students (e.g. Cambridge). Where the tourist/local visitor ratio is especially high (e.g. Canterbury), local residents are likely to become less favourably disposed towards tourism. Cities such as Venice have become almost like 'living museums', yet visitor numbers show no sign of declining. Local residents often choose to move out of the city centre rather than become part of the tourist attraction themselves. Addressing the problems of seasonality can sometimes help to ease congestion during certain periods of the year. Special events, exhibitions, or the development of business and conference tourism can help with this. However, developing non-seasonal tourism can mean that local residents never have a break from tourists, which is equally contentious.

Visitor management can be controlled through the creation of tourist 'gateways' and information centres that help to control visitor flows through the provision of selective information. Guiding services can also help to steer visitors in a certain direction so as to minimise congestion at major sites. Timed ticketing can also be appropriate in some cases as a way of limiting visitor numbers to an attraction at any one time. Self-directed or guided heritage trails or walks are also being used to encourage tourists to consider alternative routes and attractions.

Historic towns that are also World Heritage Sites often attract even higher numbers of tourists because of their enhanced status and global profile (Box 6.5). Certain heritage cities that are already suffering from over-visitation, such as Prague and Warsaw, have also recently been European Cities of Culture, which is likely to have increased their popularity further.

## Box 6.5 Maritime Greenwich: World Heritage Site and historic town

Maritime Greenwich is a good example of an historic town that is also a World Heritage Site. Prior to its designation in 1997, it was perhaps considered more of an adjunct to London than a cultural destination in its own right. However, since Greenwich also hosted the infamous Millennium Dome exhibition, it has been placed firmly on the international tourist map and visitor numbers are steady.

The millennium year afforded Greenwich the perfect opportunity to give its historic centre a 'face-lift', with the creation of new visitor facilities and services (e.g. an improved and relocated TIC and interpretation centre), special shop front and pavement design schemes, and environmental and conservation initiatives both in the town centre and along the river. Visitor management has been vastly improved since the millennium year, and the provision of information, interpretation facilities and visitor services is now excellent.

However, Greenwich still suffers from many of the problems affecting historic towns as mentioned above. It has a small historic centre which is unable to cope with high visitor numbers at certain times of the year. The heritage sites become overly congested and conservation becomes a major issue. Although some of the sites have introduced joint tickets to encourage tourists to consider lesser-known attractions, the main sites such as the National Maritime Museum, Cutty Sark and Observatory still tend to receive the most visitors. However, heritage trails and guided walks have been developed in an attempt to manage visitor flows, and a new tourist gateway is being planned to offer an alternative entry point.

Transport is a major concern, and part-pedestrianisation has resulted in local residents becoming disgruntled over the congestion that has ensued in peripheral areas as a result. However, the steady stream of traffic through the town centre still adversely affects the historic buildings because of pollution and erosion, and as a result the town centre is less attractive and less safe for visitors and residents than it might be. Car-parking provision is limited and coach dropping-off points are not always convenient. A car-free day has been

continued

piloted to give a taste of what life could be like in pedestrianised Greenwich, but this issue is still highly controversial at a number of levels.

Although Greenwich has not been keen to capitalise on its World Heritage Site status in its marketing campaigns, it has managed to create a strong brand image as 'The Home of Time' and using its maritime history. Of course, the ill-fated Millennium Dome brought the international media spotlight to Greenwich, but it appears to have done little damage to the town's reputation as a heritage destination.

Future developments are likely to focus on increasing co-ordination between key sectors responsible for the management of the town, such as the local government agencies, urban development and regeneration agencies, Greenwich University (which now occupies part of the Royal Naval College), resident communities, and local businesses. Although World Heritage Site status and the millennium year have acted as important catalysts for sustainable development and regeneration, Greenwich arguably still has a long way to go in terms of establishing itself as not just an historic town in its own right, but also as a world-class cultural tourism destination.

## Conclusion

This chapter has provided an overview of some of the key issues in heritage tourism, in particular the management of historic attractions, monuments and towns. It is clear that the future of sustainable cultural tourism is dependent on a number of important factors, most of which rely on the increasing collaboration between relevant agencies and representatives, including resident communities. Heritage management is not concerned simply with conserving the past. It has become a much more dynamic and forward-thinking process, with emphasis being placed on future as well as past generations. This includes increasing understanding and access to heritage, as well as enhancing a sense of local, national and global pride in and empathy with that heritage.

# 7 Indigenous cultural tourism

> In view of the interrelationship between the natural environment and its sustainable development and the cultural, social, economic and physical well-being of Indigenous people, national and international efforts to implement environmentally sound and sustainable development should recognise, accommodate, promote, and strengthen the role of Indigenous people and their communities.
>
> (IUCN, 1993)

## Introduction

This chapter will analyse the growth of indigenous cultural tourism in a range of environments, focusing in particular on fragile or remote locations. It is clear that cultural tourists are becoming increasingly interested in the culture, traditions and lifestyles of indigenous peoples, tribal and ethnic groups. Cultural tours or treks involving visits to or overnight stays with tribal peoples or villagers are becoming more and more popular, especially in some of the emerging destinations of the world such as Southeast Asia and Central America. Those tourists who venture in search of traditional and ethnic cultures in remote locations are often motivated partly by an anthropological desire to learn more about communities under threat from global forces, but also to satisfy their need for cultural experiences of a diverse nature. The impacts of this increasingly widespread form of cultural tourism are significant, both for the communities who are the object of the tourist gaze, and for the local and national economies that stand to benefit from tourism development. The inevitable consequence of increased tourism is often the gradual erosion of the social fabric, acculturation, and irreversible destruction of natural habitats. This form of tourism can easily become a kind of cultural voyeurism in which the local indigenous population is reduced to little more than a human zoo. However, if managed carefully, indigenous cultural tourism has the potential to benefit local communities considerably. Cultural tourism may be seen as a means of increasing the profile of indigenous peoples and to bring economic benefits. It can also lead to a renewal of cultural pride and community cohesion if managed carefully. This chapter aims to discuss ways in which the potential benefits of cultural tourism can be maximised for indigenous communities.

## What is indigenous cultural tourism?

For the purposes of this chapter, indigenous cultural tourism is used as an overarching term for both ethnic and tribal tourism, and any form of tourism that involves contact with indigenous peoples or their culture. Harron and Weiler (1992) referred to ethnic tourism as involving some direct contact with the host culture and environment, perhaps including visits to native homes and villages and the observation of, or participation in, traditional activities, such as rituals, ceremonies or dances. They note that this may or may not involve face-to-face contact with indigenous people, but that the human element is of central importance to the visit. Hinch and Butler (1996: 9) define indigenous tourism as 'tourism activity in which indigenous people are directly involved either through control and/or by having their culture serve as the essence of the attraction'. The concept of 'control' is imperative for the sustainable development and management of indigenous tourism, and one which we will return to later in this chapter.

Indigenous cultural tourism usually involves visiting native or indigenous people, such as tribal groups or ethnic minorities, in their 'natural environment'. This may be an area that is a designated cultural landscape, a national park, a jungle, a desert or a mountainous region. More often than not, it will be a remote and relatively fragile location that is not easily accessible to the average tourist. Land issues have been one of the most controversial aspects of indigenous people's lives; therefore many groups have been shifted from what was once their traditional homeland. In addition, now that the frontiers of modern tourism have been pushed to the limit, no area of the world is technically out of reach, which has serious implications both for the natural environment and the fragile cultures of indigenous groups.

Many tour operators are now capitalising on the exoticism of indigenous, ethnic and tribal groups. Activities such as hilltribe, mountain or desert trekking are becoming increasingly popular. Even without face-to-face contact with indigenous groups, tourists are keen increasingly to purchase indigenous arts and crafts as souvenirs, as well as enjoying the cultural displays and performances that seem to constitute an integral part of the tourist experience.

The following list suggests a typology for indigenous cultural tourism and the kinds of activities and destinations that are becoming increasingly popular among tourists.

- Hilltribe and mountain trekking (e.g. Thailand, Vietnam, Peru, Chile, Nepal, China, India).
- Wildlife tourism and national parks (e.g. Kenya, Tanzania, South Africa, Botswana).
- Rainforest and jungle ecotours (e.g. Brazil, Ecuador, Costa Rica, Indonesia, Malaysia).
- Desert trekking (e.g. Tunisia, Morocco, Egypt, Mongolia, India, Middle East).
- Arctic and northern periphery tourism (e.g. Canada, Alaska, Scandinavia, Greenland, Iceland).
- Village tourism (e.g. Senegal, Mali, Indonesia, Malaysia, South Pacific islands).
- Cultural heritage tourism (e.g. New Zealand, Australia, North America, Hawaii).
- Arts and crafts tourism (e.g. Guatemala, Mexico, Lapland, Mali, Panama).

The profile of indigenous cultural tourists is changing rapidly. In the past, the market was composed largely of allocentric tourists; that is, adventurous or intrepid individuals seeking the unexplored and the untouched. Although many activities such as hilltribe, mountain or desert trekking are still dominated mainly by the independent backpacker market, other forms of indigenous tourism (e.g. cultural heritage, arts and crafts, and village tourism) are now starting to form part of mainstream tourism packages. In fact, wildlife tourism on indigenous tribal lands in countries such as Kenya and Tanzania has almost become a mass tourism phenomenon. The ubiquitous cultural performances, displays, and arts and crafts markets also indicate the growing significance of indigenous culture for the tourism product.

The environments in which these activities take place are clearly diverse, usually fragile and often remote. Although this chapter is concerned principally with the cultural dimension of tourism, the environmental issues are significant, since they impact greatly on the life-styles and traditions of indigenous peoples. Issues relating to landownership are central to their struggle for survival, and in many cases, tribal groups have been forcibly moved from their homeland so that a national park, hotel complex or golf-course can be developed. The environmental impacts of tourism activities such as trekking are well documented. Such environments cannot easily withstand large groups of tourists, especially as they are often home to numerous indigenous tribal groups who then become part of the tourist landscape. Ecotourism is a form of tourism that was more concerned originally with environmental than with cultural issues, but its development has inevitably encroached on the lifestyle, traditions and culture of indigenous peoples, who often reside in the visited areas (e.g. rainforests, jungles, mountain regions). It encourages the use of indigenous guides, local products and local resources. Many tour operators have jumped on to the ecotourism bandwagon in an effort to declare themselves 'green' or 'ethical'. Of course, ecotourism can be a relatively sustainable form of tourism if it is well managed and small-scale.

It is clearly difficult to generalise about the impacts of these activities on the host community of the destination. This will depend largely on the stage of tourism development, the degree of previous local exposure to external influences, the size and structure of the indigenous community, and the nature of their lifestyle, culture and traditions. Throughout this chapter a number of examples will be given to illustrate some of the impacts of tourism on different indigenous groups, as well as discussing some of the measures that have been taken to protect their interests. The following section will demonstrate why it is necessary to support and protect such groups.

## A profile of indigenous peoples

It is important to understand some of the factors that have impacted on the lifestyles and traditions of indigenous peoples in both the pre- and post-colonial eras. The ongoing plight of indigenous minorities is more often than not a consequence of the colonial process whereby many communities were subjugated, disempowered or persecuted. European colonisers tended to assume an innate superiority over native people, many of whom had lived on the land for thousands of years before their arrival. This made them feel entitled to the best land, the right to enslave the local people, or to convert them to a more 'civilised' way of life or religion. Land was often taken from the native peoples by force. Despite their valiant struggles they were usually outnumbered, and sometimes subjected to intense violence. Colonisation also brought new and dangerous diseases such as smallpox which wiped out whole families. Indigenous culture and religious beliefs were usually suppressed, and people's children were sometimes taken away from their families by force to be educated by the white colonisers or to be married to a non-indigenous person in order to 'purify' the race.

In more recent decades, other factors have taken their toll on the lives and traditions of indigenous peoples. Mining, deforestation, road-building and civil war are all responsible for threatening the long-term survival of tribal peoples, and many are still subjected to racism, persecution and violence. The legacy that remains for many indigenous people is more often than not one of poverty, deprivation and social exclusion. Frideres (1988) describes the culture of indigenous people as 'a culture of poverty', and gives the example of the socio-economic characteristics of indigenous Inuit people in Canada as having higher rates of unemployment, suicide and incarceration, and lower income and education levels than non-indigenous peoples. Hinch and Butler (1996) suggest that this profile is often typical of indigenous people throughout the world, and that tourism is viewed consequently as a potential means of supporting community development and enhancing socio-economic status. In some cases, cultural tourism can have a positive impact, albeit small, on the revitalisation of local indigenous culture, as the case study outlined in Box 7.1 shows.

## Box 7.1 Native American Indians, colonisation and cultural heritage tourism

It is estimated that Native American Indians were present in the USA from as far back as 200000 BC. Until the first explorers arrived in the fifteenth century and the first settlers came in the 1600s, they led a relatively peaceful life in harmony with nature and were self-sufficient. Settlers brought with them fatal diseases, enslaved, kidnapped or massacred the people, and prohibited their spiritual or religious practices. Their children were often abducted and sent away to boarding-schools hundreds of miles away. Despite their valiant revolts, the Native Indians were unable to stop their lands being taken away from them, and by 1776 they were forced to cede over 90 per cent of their ancient homelands. By 1871 this figure had risen to 99 per cent. The Indians were forced to live in reservations, often in squalid conditions, and many were filled with despair.

Throughout the next century, the Native Indians remained determined to continue the traditions of their ancestors, even though they were not always accorded the protection that they were promised by the government as a kind of compensation for their great losses. Even those who fought for the USA in the First World War and the Vietnam War were not duly rewarded. It was not until the 1980s that a revival of interest in their culture, heritage and traditions began to manifest itself, partly among the tribes themselves, and partly supported by the government and tourism development. During the 1990s, many tribes began to rediscover their heritage and to explore ways of celebrating it. The development of cultural tourism has been one of the ways in which Native Indian culture is being revived. Native American cultural centres are being developed throughout North America and parts of Canada which provide a space for exhibitions, performances or demonstrations. Tours are conducted around the reservations and gift shops sell handcrafted goods. Native American museums and cultural centres can help to reaffirm cultural identity by providing a means of preserving aspects of culture such as artefacts, languages and skills. They can also provide a venue for community education and activities. The tribal museum enables Indian people to take control of the representation and interpretation of their culture and to tell their tragic but courageous story.

Source: Adapted from Steele-Prohaska (1996) and Simpson (1996)

The Native Indian story is not unique. Many indigenous peoples suffered a similar fate, and only their courage, determination and resilience have prevented the complete annihilation of their culture and traditions. Although nothing can compensate for the atrocities of the past, such groups are now accorded more political support and there is more recognition of the plight of indigenous peoples.

## The role of charities and action groups in support of indigenous peoples

Indigenous issues are highly politicised because the majority of ethnic and tribal groups are fighting for their cultural survival in an increasingly globalised world, usually without access to adequate political or legal protection and support. It should be noted that many of the conflicts that affect indigenous peoples constitute human rights abuses. Consequently, a number of international and national organisations are working for and with indigenous peoples to protect their interests. Not all of these organisations are chiefly concerned with the impacts of tourism on the lifestyles and traditions of native peoples, but tourism is only one of the many factors that threaten the well-being and future of indigenous groups.

One of the highest profile organisations devoted to indigenous peoples is Survival for Tribal Peoples, which is the only worldwide organisation supporting tribal peoples through public campaigns. It was founded in 1969 in response to massacres, land theft and genocide in Brazil's Amazonia. Survival has supporters in eighty-two countries, and it works for tribal peoples' rights through campaigns, education and funding. This includes letter-writing, lobbying politicians, disseminating information and supplying legal advice. Tribal peoples are given a platform from which to both defend and represent themselves. Much of their work aims to contest the notion that tribal peoples are relics and that they are inevitably going to die out or be assimilated because of 'progress'. Survival is an autonomous organisation that refuses national government funding in order to ensure freedom of action. Although it is not the organisation's main preoccupation, Survival is involved in some tourism-related campaigns, often in support of Tourism Concern. Tourism Concern is another high-profile organisation which campaigns at both international and national level against the exploitation of indigenous peoples. Many of the campaigns relate to land-use conflicts, the displacement of traditional industries, water consumption, the sex tourism industry or cultural conflicts. They have also established a Fair Trade in Tourism Network which fights for equitable and ethical trading rights for disadvantaged communities in poorer regions of the world.

The American-based Rethinking Tourism Project is an indigenous non-profit organisation dedicated to the preservation and protection of lands and cultures. Much of its work is based on community education and the sharing and dissemination of information and resources about tourism. Other organisations such as the World Council of Indigenous Peoples and Cultural Survival also protect the rights of indigenous peoples. Again, much of their work is based on research, dissemination of information, campaigns, education, conferences, and, most importantly, supplying a strong support network for indigenous groups worldwide. It is worth noting that there are also many other national and regional groups that campaign on behalf of indigenous, ethnic or tribal groups, for example, in North America or Australia. Many of these organisations work closely with the international organisations, and can be accessed through linked websites.

The majority of these groups are campaigning for indigenous rights, autonomy, and the control or freedom to develop tourism in a way that befits best the needs of the local community. The following sections will explore in more detail some of the issues relating to indigenous community tourism.

## A community-based approach to indigenous cultural tourism development

The importance of a community-based approach to tourism development has clearly been recognised over the past decade, increasingly in the context of the sustainable tourism debate. Murphy (1985) argued strongly for communities to play an integral role in the development of tourism, and proposed an ecological approach which emphasised the need for community control; however, the debate still continues as to how an appropriate and sustainable form of community planning should be implemented. The following list suggests some key principles for community tourism as advocated by Tourism Concern. 'Community' in this context is defined as 'a mutually supportive, geographically specific, social unit such as a village or tribe where people identify themselves as community members and where there is usually some form of communal decision-making' (Tourism Concern/Mann, 2000: 18).

1 Community tourism should involve local people. This means that they should participate in decision-making and ownership, not just be paid a fee.

2 The local community should receive a fair share of the profits from any tourism venture.
3 Tour operators should try to work with communities rather than with individuals. Working with individuals can create divisions within a community. Where communities have representative organisations these should be consulted and their decisions respected.
4 Tourism should be environmentally sustainable. Local people must benefit and be consulted if conservation projects are to work. Tourism should not put extra pressure on scarce resources.
5 Tourism should support traditional cultures by showing respect for indigenous knowledge. Tourism can encourage people to value their own cultural heritage.
6 Operators should work with local people to minimise the harmful impacts of tourism.
7 Where appropriate, tour operators should keep groups small to minimise their cultural and environmental impact.
8 Tour operators or guides should brief tourists on what to expect, and on appropriate dress, taking photos and respecting privacy.
9 Local people should be allowed to participate in tourism with dignity and self-respect. They should not be coerced into performing inappropriate ceremonies for tourists and so on.
10 People have the right to say no to tourism. Communities which reject tourism should be left alone.

(Tourism Concern/Mann, 2000)

Consensus and control are key issues, and the political nature of the planning process continues to be a major difficulty (Hall, 2000). A pluralistic approach to community-orientated tourism planning, as advocated by Murphy (1985), assumes that all parties have an equal opportunity to participate in the political process (Hall, 1994). Jamal and Getz (1995) provide a critical analysis of collaboration and co-operation, stating that power imbalances often act as a significant barrier to successful collaboration. Reed (1997) suggests that power relations are indeed an integral element in understanding community-based tourism planning and the relative success of collaborative efforts. It is clear that few communities have equal access to political and economic resources, especially aboriginal peoples and indigenous minorities who are often politically, economically and socially disadvantaged. Community-based tourism can offer such communities the chance to move towards greater political self-determination, but only if local control is maximised. As stated by Butler and Hinch (1996: 5), tourism should be planned and managed so that 'indigenous people dictate the nature of the experience and negotiate their involvement in tourism from a position of strength'.

Indigenous groups have sometimes had no control over tourism development whatsoever, but, as discussed above, charities, action groups and other political organisations are trying to ensure that consultation and involvement are maximised. This may include an advisory role at the planning stage; joint or sole management of key tourism initiatives; employment in or ownership of tourism-related businesses. However, as outlined by Hinch and Butler (1996), many tourism ventures are dominated by non-indigenous groups with strong ties to the global tourism industry.

The case study outlined in box 7.2 discusses the potential for community-based Inuit tourism in the Canadian Arctic. It illustrates clearly the complexity of developing a form of tourism that is acceptable to the local indigenous community and which is at the same time economically viable and environmentally sustainable.

## Box 7.2 Inuit tourism in the Canadian Arctic

Tourism is largely welcomed by the governments of Arctic regions for the economic benefits it brings, especially given the decline of traditional industries such as fishing and agriculture. However, the potential environmental and socio-cultural impacts of such development need to be considered carefully, as the Arctic is populated by Inuit peoples, who also have a stake in the land's development.

Inuit people in Canada generally have higher rates of unemployment, suicide and incarceration, and lower income and education levels than non-indigenous peoples (Frideres, 1988). The migration of Inuit people to permanent communities or settlements did not take place until the 1960s, prior to which time they were largely nomadic or semi-nomadic, so few Inuit elders were born and raised in the areas in which they are living today. Permanent settlements provide communities with basic services such as a water supply, power, air transportation, education, medical services and sanitation (Addison, 1996). Despite the creation of social benefits, such as permanent housing and employment, Inuit people appear to have suffered from a loss of cultural identity as a result of being displaced in this way. Coupled with the impacts of globalisation, it is difficult for communities to retain a strong sense of their cultural heritage and past traditions.

Inuit communities are generally characterised by a 'mixed economy', but tend to rely largely on income from domestic production, such as hunting, fishing, trapping and gathering. However, little cash is generated from domestic production, and unemployment is high. The Canadian Arctic is essentially a 'welfare economy', and the Canadian government is committed to maintaining support of the Inuit. Tourism development may offer a significant economic opportunity for such communities, but care must be taken to develop this sector in a way that befits best the social and cultural structure of Inuit community life.

Previous community-based tourism development initiatives in the Canadian Arctic have tended to be largely unsuccessful and 'unsustainable' in the long term due to a lack of internal control and community consensus. Studies have suggested that there is a direct correlation between the degree of local resentment and antagonism towards tourism in the Arctic, and the level of external versus internal control (Smith, 1989; Anderson, 1991; Grekin and Milne, 1996). Much of the previous development of tourism *to* the Canadian Arctic appears to have been based on external control, whereas tourism at the destination is characterised by a high degree of local control (Butler and Hinch, 1996). This is mainly because in 1983 the Government of the Northwest Territories (GNWT) established a 'community-based' tourism strategy which aimed to develop an environmentally and culturally sustainable form of tourism that maximised economic benefits for residents as a source of income and employment (GNWT, 1983).

Tourism development projects in regions such as Pangnirtung in the Baffin region often failed to give communities adequate control over tourism, as government agencies held financial and political control over the development, and the Tourism Committee played an advisory rather than a decision-making and initiating role. Although it led to greater community involvement in economic development, it was felt that more education and management training was needed for local people, and that elders should have been more involved in the interpretation of Inuit culture and history.

Developments in Pond Inlet failed to optimise the economic benefits of tourism because of the lack of product innovation, and technological and marketing expertise on the part of both tour operators and local residents. Although nature-related activity packages and arts and crafts tourism were developed relatively successfully, there was clearly a need to diversify the product further, and to give the destination a competitive advantage. Tourism also needed to be linked to other sectors of the local economy, such as agriculture or fishing. A lack of communication and co-operation at community level was also identified as having been one of the major barriers to the successful development of tourism (Grekin and Milne, 1996).

## Levels of indigenous involvement in cultural tourism development

The extent of indigenous involvement in and control of cultural tourism development is very variable, and will depend very much on the context in which development is taking place and the degree of local support. Although indigenous peoples are still rarely given complete control or ownership of tourist sites or attractions, there have been definite moves towards consultative, joint or co-operative management. Co-operative arrangements can be highly beneficial as long as indigenous peoples are treated as equal partners. However, ultimately, the majority of indigenous peoples are likely to be seeking the kind of empowerment that enables them to move towards sole ownership and management of tourism venues and initiatives. However, better support is often needed in terms of funding, education and business skills training.

The two case studies outlined in Boxes 7.3 and 7.4 illustrate the changing nature of local participation in tourism development.

### Box 7.3 Maori participation in cultural tourism

Maori have been involved in New Zealand's tourism industry for over 140 years, and much of this development has been in the Rotorua region in North Island. Like many countries with a colonial past, the Maori and Pakeha (white settler) relationship has been somewhat fraught at times. Whereas many of the earliest tourism initiatives came from the Maori themselves (e.g. accommodation, guiding, transport), many were displaced by Pakeha in later years (Barnett, 1997).

The resurgence of interest in Maori culture over the past twenty years or so has led to the recognition of the importance for Maori control over tourism development, cultural production and the authentic representation of Maori culture. Since the 1990s, the Aotearoa Maori Tourism Federation has struggled to represent Maori interests and to protect Maori culture. Ryan (1996) comments on the difficulties of categorising Maori involvement in tourism, as practices vary between locations, industry sectors and stages of business. Emphasis still tends to be placed on cultural performances and handicrafts, whereas it should be recognised that Maori are now owners and managers of an increasing number of tourism ventures (e.g. accommodation, attractions, transport, tours).

Barnett (1997: 472) summarises the way in which the Aotearoa Maori Tourism Federation (1996) categorised the Maori tourism product:

- *Entertainment*: includes concerts performed in hotels, restaurants and museums.
- *Arts and crafts*: items produced for tourists that are generally sold in souvenir shops.
- *History and display of artefacts*: generally referring to treasures held in museums and art galleries.
- *Guided tours*: many activities included in this category (e.g. guided bush walk, half-day mini-van tour, guided tour of a *marae*.

Other Maori tourism operations included mainly accommodation-related initiatives.

Barnett (1997: 473) concludes by saying that 'the issue of control is what lies at the heart of Maori tourism development in New Zealand and is the biggest concern of many Maori'. This is true of the majority of indigenous peoples who are involved in tourism development, as emphasised earlier in this chapter.

## Box 7.4 The role of Iban longhouse people in cultural tourism development

The Iban are the largest ethnic group in Sarawak, Borneo (Malaysia), and their longhouses and villages are visited as part of river safari cruises. Iban people practise traditional customs and follow the animist religion. Some areas have been visited by tour groups since the 1960s, whereas tours to others have only begun more recently. Tourists usually spend one or two nights in longhouse accommodation and are often invited to eat with the resident hosts. Tour groups are generally quite small, and the smaller groups are clearly afforded a more intimate experience. Activities include jungle walks, cultural performances (e.g. dances, music, songs), handicrafts shopping, witnessing special events and ceremonies.

Generally, local people have provided longboat transport, and worked as jungle guides, cooks and cultural performers. However, Iban hosts are gradually moving towards being 'culture managers' rather than only 'culture providers'. Although tour operators are often responsible for the arrangements, they pay the Iban hosts for the services they provide, a process that is negotiated with the headman or tribal elder. Consultation usually takes place with a longhouse tourism committee and Iban people are allocated certain tasks or roles which are organised on a roster basis. On the whole, community tourism appears to be managed fairly and effectively. The level of Iban involvement ranges from the community acting as service supplier to tour operators; community control of tourism and accommodation facilities; or in some cases (e.g. Nanga Stamang) the formation of partnerships with ecotourism companies.

The degree of genuine or 'authentic' interaction with visitors seems to vary from village to village. Some tours focus on 'cultural sightseeing' with a minimum of interaction, whereas less formal 'meet the people' experiences afford more intimate contact. Clearly, the degree of interaction should be determined by the Iban people themselves, rather than the tour operators or tour groups.

Source: Adapted from Zeppel (1997a)

Wall (1996) identifies the following issues as being the most significant for the future involvement of indigenous peoples in the management of tourism and heritage sites:

1 The precise meaning of partnerships and shared and co-operative management.
2 The role indigenous people play in management.
3 The ownership of land.
4 Implementation of mechanisms.
5 Relevant legal, constitutional and socio-economic contexts.

As discussed earlier in the chapter, international agencies are keen to ensure that indigenous people are given the support they need to become involved at all levels of tourism development and management, but there is still a long way to go.

Various management tools are being used to protect and support traditional cultural practices, and to control the representation of indigenous culture. Smith (1997) suggests the following as a tool for the development of tribal tourism, which is particularly relevant to homogenous groups that are living in remote locations:

1 *Habitat*: access, proximity, appeal, diversity, resources, marketing.
2 *Heritage*: cultural resources, museums, interpretative centres, ceremonials, experiential, marketing.

3 *History*: culture contact, decision makers, conflict, resolution, showcase today, marginal men/women, marketing.
4 *Handicrafts*: Heritage crafts, innovation, miniaturisation, marketing.

(Smith, 1997)

This is a useful model, as it shows ways in which tribal people can develop their habitat for tourism. Accessibility of the site and its location are key considerations. There is clearly no point developing cultural tourism in a location that is inaccessible to tourists. Equally, tourists have to know that the site exists, and therefore marketing is a major consideration. The interpretation of history and heritage should be undertaken by the people themselves in ways that they consider to be appropriate. The establishment of museums, cultural centres and performance space is a good way of exhibiting local culture. The development of arts and crafts initiatives is also an increasingly popular way of maximising economic benefits for communities. The following sections will explore some of these issues in more depth.

## The cultural representation of indigenous peoples

Chapter 6 placed emphasis on the need for a more inclusive and democratic politics of representation. As stated by Ali (1991: 211):

> The 'struggle over the relations of representation' though it is not yet over, echoes strongly the legacy of empire in its 'us' and 'them' colonial relations. A politics of representation is altogether a more complex, more interesting and more open challenge for the future.

Museums and galleries have been subject to criticism in recent years because of their inadequate or inaccurate representation or interpretation of black, indigenous and minority cultures in exhibitions. As stated by Creamer (1990: 132), 'a white, Western, colonizing ideology has provided the intellectual framework for interpreting indigenous cultures the world over'. Simpson refers to the concept of 'scientific colonialism' whereby anthropologists have claimed that cultural colonialism continues to control the representation of indigenous arts and culture, especially in museum collections and exhibitions:

> Indigenous groups in Australia, New Zealand and North America, while unique and diverse, share certain similarities with regards to their treatment historically at the hands of colonial powers, their status as disempowered minority communities in their own countries, and the history of their treatment by anthropologists and museum interpreters.
>
> (Simpson, 1996: 80)

Ethnic minorities tend to be under-represented in the arts and museum world, as they have traditionally lacked the power and control to determine exhibition content and interpretation. Hence, interpretation of indigenous collections or exhibitions is often left in the hands of non-indigenous peoples who may or may not understand fully the culture and traditions they are depicting. Traditionally, the culture of indigenous peoples has been fossilised in museum exhibitions or viewed with nostalgia, implying that it has vanished or disappeared, rather than being dynamic and ongoing. As stated by Dann (1996b: 366), 'museums emerged as warehouses of assembled artefacts rather than representations of living cultures'. However, museum exhibitions are now focusing increasingly on the 'truth' of indigenous and colonial history, as well as attempting to represent and interpret indigenous traditions and culture more accurately.

Lifestyles and traditions have often been romanticised or described as 'exotic'. Jordan and Weedon (1995: 489) describe how Australian Aborigines are typically perceived in the West, a description that could apply easily to many other indigenous groups:

> Most images in the West of Australian Aborigines are of 'primitive', 'tribal' people – the uncivilised 'dying race' of the anthropologist, *National Geographic* and documentary films; the dark-skinned savages of cultural evolutionist texts; the scary nature-people of 'Crocodile Dundee'; the 'stone age artists' of Australian travel brochures. The images are of naked primitives with their boomerangs, stone axes and spears, their ancestral sites and secret rituals, their traditional songs, dances and Dreamtime stories, their bark paintings and body decorations – living in harmony with nature, blending into the flora and fauna.

They remark that this is clearly a romanticised vision of the exotic native of the white Western imagination. It stands in stark contrast to the reality of the large groups of uprooted, oppressed and demoralised Aborigines who are living in shanty towns plagued by poverty and alcoholism.

Whittaker (2000) comments on the way that postcards over the past century have traditionally depicted indigenous people in a stereotyped, exoticised or eroticised way. The patronisingly benevolent and covertly racist depiction of the 'noble savage' is perhaps the best-known example of this. Most postcard production has pandered traditionally to a white, Eurocentric mythology of the 'Other'. Because indigenous peoples were perceived increasingly as a dying breed, the capturing and fossilising of cultural images was deemed imperative:

> The world's indigenous people provided a limitless reservoir for a traffic in images. The myth of the time was that there is a world of types, a veritable smorgasbord of racial, ethnic and indigenous people. Deeply-rooted beliefs about colonial entitlements, sanctified by science and its closest lieutenants, truth and reality, kept these types prisoners of the roving camera. The task of scientist and layperson alike was to document and preserve the huge variety of peoples, surely a global mission to record each and every 'type' and 'species' and 'culture'.
>
> (Whittaker, 2000: 428)

More recent depictions of indigenous peoples in tourist postcards tend to focus on traditional (often rapidly disappearing) practices and industries (e.g. fishing, weaving, farming). Native dress, arts and crafts, and cultural performances also tend to feature in order to add 'local colour'. However, Whittaker (2000) ends by commenting on the way in which postmodernism has led to the reclaiming of indigenous culture by the people themselves who are now often producing their own postcards, and hence controlling representation.

Dann (1996b) comments on the way in which the 'Other' has been traditionally depicted in travelogues. Images of natives are often overtly seductive, exotic, quaint or cute relative to the traveller's own culture. Although little interaction is depicted between host and guest, locals are usually depicted as being in a subordinate position. In some cases local people are shown to be exploitative, hostile or something of a nuisance; in other cases they are marginalised or deemed irrelevant. Dann (1996b: 366) comments on this 'new' form of tourism: 'Even in the very act of sightseeing, attractions become ways of remaining out of contact with destination peoples.'

However, at the other end of the spectrum are those tourists who are keen to spend as much time as possible in close contact with local people and to experience authentic culture. Increasingly, images of native peoples are being used to attract tourists.

Kirschenblatt-Gimblett (1998) notes, for example, that Australia and New Zealand tend to identify their uniqueness as tourist destinations with the indigenous, a fact that is becoming increasingly apparent in their national marketing campaigns. Similarly, Power (1997: 54) emphasises the significance of aboriginal arts to the cultural tourism product in Australia:

> Today it is the visual arts that offer Aboriginal Australia its greatest empowerment in our efforts to have our culture recognised locally, nationally and internationally. Foreign statistics reveal that at least 60% of all visitors to Australia cite an interest in Aboriginal culture as one of the reasons they come down under.

The growing interest in indigenous culture and arts and crafts tourism raises important questions about concepts such as authenticity and commodification. The performing arts are increasingly forming an integral part of the indigenous cultural tourism experience. This may be in the form of music, singing, dancing, festivals or rituals. Local festivals in particular are a big draw, especially in countries such as India which are renowned for their colourful celebrations of life, religion and spirituality. More often than not, they represent an authentic cultural experience for the visitor, as they are neither staged nor adapted to suit tourists' tastes. However, the same cannot be said for many other kinds of cultural performance and display. The growing interest in indigenous arts and culture can help to support cultural continuity or the revival of traditions; however, it can also compromise the authenticity of the art form:

> Whenever tourism becomes an important component of the local economy there is an increase in interest in native arts and crafts. However, it is the cultural components that have value to the tourists that have been preserved or rejuvenated and not necessarily those which are highly valued by the local culture. This type of cultural awakening has sometimes made host populations more aware of the historic and cultural continuity of their communities and this may be an enriching experience. In other cases the new appreciation of indigenous culture, the revival of ancient festivals and the restoration of cultural landmarks have emerged in ways which pose long-term threats to the existence of culture in its original form.
>
> (Mathieson and Wall, 1992: 175)

Turner and Ash (1975) suggest that the majority of tourists have a fairly limited sensual and aesthetic sense; therefore indigenous culture sometimes needs to be presented to them in a simplified format which then compromises the art form. The implication here is that tourists are not apparently 'sophisticated' enough to appreciate the more complex art forms of indigenous communities (they refer to the example of Balinese culture and art). Many tourists do want simply a taste of local culture, not a lengthy performance of dancing, singing or music which requires considerable concentration or phenomenal endurance. For example, the Kathakali story-plays in southern India would traditionally be performed all night for a Hindu audience in a temple. However, most tourists are quite content with a performance lasting one-and-a-half hours in a theatre or cultural centre. The same is true of some Maori performances. Tahana and Oppermann (1998) note, for example, that Maori cultural shows tend to be modified to suit the needs of the clientele in hotels, unlike the more traditional performances in the *marae*. This does not necessarily make the experience a simplified or 'inauthentic' one, nor does it have to compromise the integrity of the artists. The cultural experience is simply condensed. Tourists rarely understand local dialects; therefore translations might be provided of songs or plays, or some kind of interpretation offered. This is surely educational and informative rather than being an over-simplification. It is unrealistic to expect the majority of tourists to gain an in-depth understanding of indigenous cultural and art forms in the short time available to them, especially given that they may be experiencing a whole host of other attractions and events. This does not imply that they are necessarily superficial and gullible. This depends very much on the type of tourist, the context in which they find themselves, the kind of experience they are expecting, and the kind of product that is being offered to them.

It is assumed generally that cultural tourists want to experience the authentic culture of the local environment and its people, and will go to great lengths in search of the ultimate authentic experience. This may involve pushing the frontiers of tourism further and further, and visiting the most remote locations. Selwyn (1996) explores the concept of the tension between the quest for the 'authentic Other' and the quest for the 'authentic Self'. It is not unusual to hear of tourists who want to spend time with native peoples in an attempt to 'find themselves'! This is particularly true of the hippy tourists who first went to countries like India in the 1960s and 1970s. However, it is evident that some tourists are not so much seeking authenticity, but are actually in search of a romanticised vision of 'primitive' or native living. Those who are keen to escape the trappings of the globalised, material world will relish the notion of spending some time in a remote village without running water and electricity. Ironically, the villagers themselves might be just as keen to procure Western goods such as Coca-Cola, Nike trainers and satellite TV, as the case study outlined in Box 7.5 demonstrates.

## Box 7.5 Hilltribe trekking in Thailand

Tourists are increasingly being encouraged to take organised treks into the hilltribe country of northern Thailand in the region of Chiang Mai and Chiang Rai. Most of the trips involve a two- or three-day trek with a local guide through jungle areas and include overnight stays in tribal villages. The activities offered are mainly hiking, bamboo rafting and elephant riding. The tribal groups visited depends on the routes taken, but it is the usual practice to visit at least three different tribal groups. There are at least seven or eight different indigenous groups living within this area of Thailand. Many are subsistence farmers (e.g. rice, corn, livestock, opium), and some are involved in local crafts production (e.g. textiles, embroidery, silversmithing). Most groups have their own dialect, and the majority practise the animist religion. Opium addiction is common, and this is not helped by the fact that some tourists view opium-smoking as one of the highlights of the trip, especially as many of the guides are also clearly addicted. Groups tend to be fairly small (around ten people), but it is not uncommon to meet other groups of tourists during the trek, although usually not within the same tribal village.

It is difficult to adjust to the idea of being a cultural voyeur, and there is something a little intrusive about observing people going about their daily business, washing, cooking or farming. The attitudes of the villagers also vary enormously from one village to the next. In some cases the villagers have become fairly indifferent to tourists, even the younger children, who have clearly seen one tour group after another arrive and depart from their village. In others, where there has been less exposure to tourism, the villagers seem to be enthusiastic about the visit, and might entertain the tourists with dancing or singing in the evenings. Much of the time this appears to be a spontaneous and authentic gesture rather than a staged event. In all of the villages without exception, following the evening meal, the villagers will descend on the tour group with their handicrafts and will urge you to buy. At first, it seems only courteous to purchase something, however small. Nevertheless, it can be rather disconcerting to discover the 'Made in Bangkok' label on a supposedly authentic souvenir!

continued

**Plate 7.1** Backpacker gets to know Thai hilltribe village children during trek

Source: Author

Tourists sleep in wooden huts which are fairly rudimentary, they make use of pit toilets, and they bathe in the local river. There are few Western-style amenities apart from the ubiquitous bottles of Coca-Cola, which invariably find their way into the remote jungle areas! Tourists assume that they are 'going native', staying in accommodation that is typical of the local style, eating basic foods, and are cloistered from the trappings of the material world. It is only when you venture further afield that you realise that some villagers are actually living in brick houses down the road with satellite television and well-stocked local grocery stores! Tourists in search of an authentic encounter with indigenous peoples may well feel deceived and disappointed. However, in some ways, it is amusing and reassuring that many of the villagers are somehow having the last laugh. While Western tourists are clamouring to 'get back to nature', the villagers are increasingly enjoying the comforts of Western-style living, and who can blame them?

## Indigenous arts and crafts tourism

Richards (1999b) describes how culture, crafts and tourism are becoming inseparable partners. Arts and crafts production can play a central role in the local economy of indigenous communities, and tourism can help to support and strengthen the continuation of local cultural production. The kinds of souvenirs that might be purchased depend very much on local resources and traditions, but may include textiles, carvings, pottery, paintings and jewellery, among others.

It is noted that the authenticity of arts and crafts is considered to be very important for the majority of cultural tourists. Tourists want to be assured that the product they are buying is made by a local craftsperson, and reflects traditional methods and a design which is characteristic of the local area. Of course, the commodification and mass production of tourist souvenirs is widespread, but many craftspeople are now using government-approved stamps of authenticity to protect local production and to reassure tourists seeking authentic products.

Nevertheless, the proliferation of pseudo-traditional art forms is sometimes a cause for concern, especially among indigenous peoples who are keen to preserve their traditions. Local arts and crafts production is increasingly under threat in many parts of the world, and although tourism can lead to a revival of traditions that might otherwise be lost, numerous examples of cultural change are inherent in the art forms that are produced. Graburn's (1976) work on the changes in Fourth World arts and the commercialisation of cultural traditions is still one of the most comprehensive studies on this subject. Schadler's (1979) work on arts and crafts in Africa is also worth consulting, as he outlines many of the modifications that have been made to traditional and tribal art forms in response to the demands of tourism. More recently, Mathieson and Wall (1992) cite numerous examples of the changes that have taken place within traditional art forms over time. The following shows the three major phases of change in traditional art forms resulting from outside contact:

1 The disappearance of traditional artistic designs and art and craft forms, particularly those with deep religious and mythical affiliations. This is followed by;
2 The growth of a degenerate, unsophisticated replacement which develops in association with mass production techniques. This is often followed by;
3 The resurgence of skilful craftsmanship and distinctive styles incorporating the deeper cultural beliefs of the host society. This phase is a response to the deleterious impacts evident in Phase 2 and also to the gradual decline in the symbolic meaning of traditional arts which also occurs in the second phase.

(Mathieson and Wall, 1992: 165)

This is a largely positive model, implying that the final outcome leads to a resurgence of interest in traditional designs and production methods, although there are still concerns about changing production methods, the quality and authenticity of the products, and the nature of the tourism market. However, De Vidas (1995: 81) makes the point that: 'Instead of dwelling on the dialectic of the maleficent or beneficent effects of tourism on visited societies it will be more pertinent to focus upon their social, economic and cultural reorganization in response to that access.' Indeed, more recent studies of arts and crafts tourism have tended to focus on the needs of local artists, producers and craftspeople and the protection of their interests. As the demand for crafts tourism grows, the role of the middleman is becoming problematic for the distribution and sales of arts and crafts products, as it can easily lead to the exploitation of local people. Fair trade initiatives (e.g. Tourism Concern's Fair Trade in Tourism Network) are therefore being set up in order to protect local producers from exploitation, and more emphasis is being placed on local training needs.

## Conclusion

This chapter has provided an overview of the post-colonial status of indigenous peoples in a range of environments. Clearly, this account is by no means definitive, but it is evident that the majority of tribal and ethnic people have suffered a similar fate. However, over the past few decades, indigenous peoples have increasingly been afforded the kind of political, legal and economic support that is imperative to their future survival. Rather than being perceived as a dying species, there is a growing appreciation of their resilient and dynamic culture and tradition, and a recognition of the need to protect it. Although land-use remains perhaps the most controversial and unresolved issue, a great deal of progress has been made in other areas of indigenous development. Tourism may be viewed as one of the most positive forces for change in terms of economic benefits, conservation measures, and the protection or revitalisation of cultural traditions. The key concepts of local empowerment, self-determination and control need to be adhered to, but if tourism is managed responsibly and ethically, its contribution to the cultural survival of indigenous peoples can be invaluable.

# 8 The arts, festivals and cultural tourism

> To the world of tourism, the arts bring style, culture, beauty and a sense of continuity of living.
>
> (Zeppel and Hall, 1991: 29)

> Ethnic minority arts have a vast amount to give to the general cultural life of this country. Not only can they provide . . . 'colour' . . . but also new cultural forms and – in the case of folk art – a refreshing link with a rural tradition. For members of the minorities themselves, their cultural traditions represent their own identity.
>
> (Khan, 1976: 133)

## Introduction

The aim of this chapter is to analyse the relationship between the arts and cultural tourism, which has arguably been somewhat fraught in the past, with rather different philosophies governing the two sectors. Until relatively recently, there appeared to be a general lack of communication between arts and tourism organisations and a poor understanding of respective activities, priorities and objectives. Arts organisations were not always well informed of tourism trends and markets, just as tourism organisations sometimes failed to understand the needs of the arts. However, both the visual and performing arts sectors are increasingly providing visitors or tourists to a destination with a whole range of colourful exhibitions or spectacles to observe, admire or participate in. This is true particularly of destinations where vibrant local cultural events such as festivals or carnivals draw in the crowds. Many such events serve as an assertion or expression of the culture and identity of minority groups, such as ethnic communities. There are therefore a number of sensitive issues surrounding the presentation of such events, and the interrelationship between performers and their audience. This chapter will focus predominantly on the importance of ethnic and minority arts to the cultural tourism industry; therefore the historical and political backdrop of post-colonialism and emergent multiculturalism will serve as the overarching theoretical framework. There will also be some analysis of changing perceptions of art and its relationship to more democratic interpretations of culture.

## Democratising the arts: values, policies and politics

The increasing democratisation of the arts over the past few decades has emerged as a response to a perceived elitism and inherent racism within cultural policy-making in many Western societies. Before the decolonisation of Western empires, the major preoccupation within cultural studies, particularly in the UK, was with class issues and the value of so-called mass or popular culture. As discussed earlier in this book, theorists such as Raymond Williams challenged the prevailing notion that so-called 'high culture' was in some way superior to the culture of the masses. However, since decolonisation, emphasis has been placed increasingly on issues of cultural diversity, diaspora, expressions of identity and hybridisation.

At this juncture, it would perhaps be of some interest to discuss the relationship between the arts and culture, as the terms are not necessarily synonymous, although Lippard (1990: 14) suggests that they are inextricably linked: 'When culture is perceived as the entire fabric of life – including the arts with dress, speech, social customs, decoration, food – one begins to see art itself differently.' Proponents of the 'Culture and Civilisation' movement of the nineteenth century argued that Culture (with a capital 'C') should be ennobling or civilising, leading to personal advancement or enlightenment. Although cultural critics like Matthew Arnold declared that this cultural enlightenment would transcend social divisions such as class, gender, religion or ethnicity, Jordan and Weedon (1995) argue that the basic tenets of such Liberal Humanist theories were quintessentially Eurocentric. Not only did they advocate the high culture and grand narratives of white, European males, but they took little account of the barriers to access to cultural education and the accumulation of cultural competence required to appreciate such dominant cultural forms. Although in countries like Britain a form of liberal, mass education had developed towards the end of the nineteenth century, the definitions of what constituted 'worthwhile' culture were (and still are) largely tenuous and essentially elitist. As argued by Bourdieu, 'artistic value' is normally placed upon forms or objects for which a high degree of cultural competence is required. Artistic value systems therefore become somewhat arbitrary.

There has been much debate about whether culture should be defined as a 'whole way of life' or as 'the arts and learning'. Williams made a case for both definitions having equal validity. Problems arise, however, in a climate of limited support and resources for culture, when decisions have to be made about whose culture should be funded; whose culture should be given a space or a voice; and how those cultures should be interpreted or represented. These decisions are fundamentally political, but they are also commercially driven increasingly by the demands of the marketplace and the power of the consumer.

If we consider for a moment the concept of art as opposed to culture, we are dealing predominantly with the works of creative expression of an artist or artists, rather than the whole way of life of a people. Lewis (1990: 5) suggests that although definitions of art are more or less arbitrary, it makes sense to use a functional definition of art as:

> a cultural practice that involves the creation of a specific and definable object – a play, a video, a piece of music for example. The function of that object is as a self-conscious, personal or collective expression of something. This distinguishes bingo from ballet.

There have been numerous fierce debates about what constitutes valuable or worthwhile art. For example, controversies abounded over Damien Hirst's sheep in formaldehyde and Tracey Emin's bed. Both pieces of art were, in accordance with Lewis' (1990) definition, a personal and creative expression of something by the artist in question. He also notes that (in accordance with the Surrealists) art generally becomes art when an artist says it is! Yet, the general public railed against this apparent insult to what they deemed to be 'art' in much the same way as the Impressionists faced public derision in the late nineteenth century. Now, of course, Impressionist art is some of the most publicly and commercially valued of any artists in the history of Western painting. However, the image of the starving artist in the attic, which characterised artists like Van Gogh during his lifetime, has never afflicted the likes of Damien Hirst or Tracey Emin, the market value of whose art is astronomically high. This is a sad irony for the artists of the past, but arguably a coup for the artists of the present. One gains the impression that Damien Hirst has had the last laugh when it comes to the apparent 'commercialisation' of his art. In many ways, his attitude resembles that of Andy Warhol who was one of the first artists to blatantly fuse art and commerce, aestheticising everyday household objects (e.g. soup cans) and drawing on popular cultural influences like Hollywood to inspire his art. Other pop artists have been similarly influenced by cartoons, comic strips and advertising (e.g. Lichtenstein, Rauschenberg). But the relationship between artist and commodity culture is arguably always somewhat uneasy, especially where the needs or desires of the marketplace start to dictate the nature of cultural production. Rockwell (1999: 94) suggests that what we may end up with is 'demotic desolation, with mass taste, whipped along by overpowering commercial interests, nearly obliterating high art'. But he emphasises again their interdependence:

> To separate the older 'high' arts from the younger mass arts denies both their best chance for healthy growth. The high arts lose vitality, while American demotic culture, writhing and heaving with mindless bestial energy, is cut loose from refining guidance.
>
> (Ibid.: 101)

There was clearly a fear in the USA that popular culture would somehow absorb high art, a feeling that was intensified in the 1950s. The idea was developed that the arts should be useful rather than beautiful, hence the shift from 'auratic' art to art being mass produced or used as a tool (perhaps for community development or regeneration, for example). Benjamin contested that the 'aura' of a work of art, which was linked to its authenticity and ritual aspects, would effectively be destroyed by mechanical reproduction. The extent to which this compromises artistic integrity is debatable. The reproducibility of the arts is perhaps harmless in the case of CDs or books, but the endless reproductions of great masterpieces as cheap prints or souvenirs has arguably detracted from the appeal of the originals somewhat.

It is difficult to conclude how far art appears to be defined increasingly by its market value rather than by its aesthetic appeal, or indeed, whether this matters. The concept of aesthetics has always been philosophically problematic. For example, although philosophers like Kant argued in favour of a universal or objective notion of aesthetics, it is difficult to define or describe this apparent 'truth'. Therefore, aesthetics is regarded more generally as a subjective notion, hence any aesthetic judgements made on pieces of art will be biased according to the eye of the beholder. Of course, it has not helped that in the history of Western art, the beholder has usually been a white, middle-class male.

## Re-dressing the balance: multicultural policies for multi-ethnic societies?

The same Eurocentric and ethnocentric biases as described in the context of art have also extended to broader cultural policy-making; thus minority groups – especially ethnic minorities – have rarely been given the chance to represent themselves or their interests adequately. This is not unique to post-imperial societies where immigration from the former colonies has been widespread, but it also affects countries such as America or Australia where the 'first peoples' have been similarly marginalised. For example, as noted by Lippard (1990: 6):

> Ironically, the last to receive commercial and institutional attention in the urban artworld have been the 'first Americans', whose land and art have both been

> colonized and excluded from the realms of 'high art', despite their cultures' profound contributions to it.

Many governments are attempting to redress this imbalance, but there is still a great deal of tokenism in some of these gestures. Mowitt (2001: 8) suggests cynically that 'In a global white world, a little local color goes a long way'. However, Cantor suggests that the US government has attempted to be highly democratic in its approach to the arts, adopting a postmodern approach to cultural policy-making:

> Postmodernism is the first form of art that accepts the idea of the radical historicity of all art. No aesthetic principles are simply true or universal; all are time-bound and culturally limited. We might even describe postmodernism as historicism elevated into an aesthetic principle.
>
> (Cantor, 1999: 174)

However, Lippard notes that one of the problems with discourse within art is that it tends to be *about* the Other rather than *by* the Other. She prefers therefore to talk of 'a common anotherness' (1990: 6) which is more inclusive of both subject and object. She also raises the problem of 'quality' which has been used as a subjective benchmark against which to measure the value of art. Quality is, of course, a subjective and relative concept in the same way as aesthetics. Brustein suggests that the NEA (National Endowment for the Arts) in the USA over-democratised the arts in a vain attempt to be more inclusive, and compromised artistic standards in the process:

> It may be just another form of plantation paternalism to bestow largesse on minority artists regardless of their abilities, to use different yardsticks for artists of color, to reward good intentions rather than actual achievements, for self-esteem is rarely achieved by means of abandoned standards.
>
> (Brustein, 1999: 22)

However, the dubious idea of 'standards' again implies some kind of subjective attribution, and diversification and democracy are not, of course, necessarily synonymous with the decline of standards.

There is clearly a need for the recognition of multi-vocal arts, but the way in which these are articulated may also need to be analysed. For example, Nonini (1999: 158) states that:

> Although those in the 'lower order' articulate their 'experience', often in critical expressive genres (such as, originally, blues or reggae music for North American blacks or Afro-Caribbean groups), much of the time they are constrained to do so by appropriating the authoritative discourse of the more powerful, at times parroting it, or at best – through some forms of irony – turning against their oppressors in petty forms of 'everyday resistance'.

Even once they have found a voice and claimed a space, many minority artists may need support in casting off the shackles of their oppressive past and finding new forms of expression and appreciative new audiences. Cultural tourism can partly afford them this opportunity, as the remainder of this chapter will demonstrate.

## The relationship between cultural tourism and the arts

The arts represent an extremely important component of the tourism product. Zeppel and Hall (1992) described arts tourism as being based on a broad array of activities, including paintings, sculpture, theatre, and other forms of creative expression, such as festivals and events. Arts tourism tends to be experiential, whereby tourists become involved in and stimulated by the activities that are presented to them.

Past research in cultural tourism studies has often focused on the relationship between tourism and heritage management. Arts tourism has tended to receive less attention, with the exception perhaps of festivals and events. Past studies often focused on the tensions or conflicts between the two industries (e.g. Tighe, 1986; Turner, 1992; Varlow, 1995). For example, Turner (1992) suggested that there tended to be a lack of understanding of respective priorities of the arts and tourism industries in the past, especially with regard to the concept of entertainment. Whereas tourism is often viewed as being focused on merely entertaining or amusing visitors, the arts are perceived as having an apparently more cerebral or educational role. He is also slightly cynical about the motives of some organisations for seeking financial support through tourism: 'there are those who suggest that "entertainment" only becomes "the arts" when it needs a subsidy' (Turner, 1992: 109). However, he does also recognise the potential benefits of collaboration, as do Zeppel and Hall (1991: 29), who state that:

> In commercial terms, the arts revitalize the tourism product, sharpen its market appeal, give new meaning to national character, and permit much tighter sales and promotional efforts. Simply stated, the arts, as an element of tourism, improve the product and strengthen its appeal, making tourism saleable.

Tourism is also important for the arts in the sense that it generates substantial revenue in terms of attendance figures and ticket sales at events and attractions, and museums and galleries are often heavily dependent on financial support from tourists. In addition, tourism can broaden the market for the arts, and increased publicity can lead to the possibility of sponsorship opportunities, which are becoming increasingly important in a climate of waning financial support.

It is perhaps useful to summarise some of the main synergies and conflicts within arts tourism development (see Table 8.1). Some of these points need little explanation. As stated by Myerscough (1988), the arts can act as a magnet for destinations, encouraging people to stay and spend money in the local economy. Tourism can help to broaden and diversify arts markets, and to raise the profile of lesser-known events. Whereas many arts events tend to attract a local audience, tourism tends to be national or international in its scope. Many arts venues and events are becoming increasingly reliant on commercial sponsorship as public sector support declines. Consequently, new sources of revenue are always being sought. However, arts organisations are often wary of diversifying audiences too much, fearing that tourism may in some way compromise the artistic integrity or authenticity of the event. The programming of arts schedules is often not compatible with the more mainstream interests of tourists; therefore arts organisations are sometimes forced to compromise to please their target audiences. Arts organisations are perhaps right to be sceptical in such cases, as it means the potential dilution, commercialisation or trivialisation of art forms.

A number of recent conferences and publications have analysed the relationship between the arts and tourism in more depth, focusing mainly on the potential benefits of collaboration, but also accepting the limitations. Hadley (1999: 8) provides an overview of the current relationship between the two sectors, concluding that increased dialogue is needed in order to move forward in terms of policy-making and developing joint initiatives:

**Table 8.1** ***Arts tourism: synergies and conflicts***

| *Benefits of arts tourism* | *Tensions in arts tourism* |
|---|---|
| • Tourism can generate revenue through ticket sales at events and attractions<br>• Tourism can broaden the market for the arts<br>• Increased publicity can encourage sponsorship<br>• Arts can enhance the image of a destination<br>• The arts can boost the evening economy and encourage overnight stays<br>• Arts tourism can play a key role in regeneration | • The arts still have a relatively low profile in the tourism sector<br>• Priorities in the arts and tourism sectors are often very different<br>• Arts organisations can be sceptical about the practical and financial benefits of tourism<br>• Competing marketing priorities exist (e.g. between local and non-local audiences)<br>• Inadequate lead times for some exhibitions or events<br>• Constraints imposed by the nature of the arts product (i.e. in terms of programming, artistic integrity, etc.) |

> Like any relationship in trouble, the causes of the trouble are deep-seated and long-standing. No easy answers therefore – just a will to do things differently in future, a change of attitude, greater tolerance, a commitment to finding out what the other side needs – and isn't getting.

Hughes (2000) discusses the relationship between the arts, entertainment and tourism and provides some useful and comprehensive definitions of cultural and arts tourism, and an in-depth analysis of audience participation and the potential for audience development through tourism. He recognises the importance of developing arts tourism, but he is also realistic about the limitations. He suggests, for example, that the impact of the arts on local economies is often exaggerated, and that developing arts-related tourism is not always in the best interests of the arts: 'It may be naïve to assume that commercial organizations will forgo revenue and profit in the cause of artistic integrity and creativity and of stimulating new, experimental and minority-appeal artistic creations' (p.197).

In destinations where arts tourism development is flourishing, tourism organisations tend to view the arts as an attractive way of boosting the cultural tourism product. Art museums and galleries are a big draw for tourists, particularly in large cities. For example, there appear to be a number of 'must-see' international art museums and galleries, particularly in Europe, such as the Louvre in Paris, the National Gallery in London, the Prado in Madrid and the Uffizi in Florence. These are clearly unique collections of some of the world's most beautiful and valuable pieces which can be seen only in those cities. In contrast, the performing arts are often more global in the sense that opera, ballet, classical music, theatre plays and musicals tend to be moveable feasts, and can often be viewed in the tourists' own country or city. The increasing globalisation of the arts, particularly in the urban context, is an interesting phenomenon. Many art forms that started out as small-scale, local traditions have now become globally available and universally popular. Flamenco dancing is a good example. Many community festivals or events have also started to attract national and international audiences, despite having traditional roots. Some examples of these will be given later in the chapter.

## Participation in arts tourism

It is generally recognised that arts tourists tend to be the kind of people who regularly visit arts attractions or venues at home (Richards, 1996; Hughes, 2000). The profile of arts tourists tends to be broadly similar to that of more general cultural tourists. They usually have relatively high levels of education, income and cultural competence. However, participation in the arts is a hotly debated issue in arts management literature. For example, it has long been recognised that the contemporary arts scene in Britain tends to be dominated by predominantly white, middle-class, middle-aged audiences. In addition, minority and ethnic arts activities tend to be under-funded and less supported than mainstream arts. There is still arguably a certain snob value attached to the arts which is linked partly to the nature of so-called high art forms, but also to the host institution, its location and its pricing structure. For example, many large arts venues such as Covent Garden Opera House in London have been criticised in the past for not facilitating access to a broader range of potential audiences. Barriers to access clearly need to be overcome before the arts can truly become democratic in their audience development, and funding and support need to be increased for minority activities.

In terms of motivation, the high arts (e.g. opera, ballet, classical concerts) often tend to attract audiences who are motivated partly by the prestige value or social status of attending such a performance (Dimaggio and Useem, 1978; Zeppel and Hall, 1991). Compare this with the genuine and spontaneous delight that spectators and participants often take in a festival, carnival or rock concert, and it is not difficult to see why certain arts events are more popular with tourists and the general public than others. They are more inclusive, since barriers to access such as the need to accumulate cultural capital, a better education or a higher income level have been removed. In addition, many such events are staged or performed by local, often ethnic artists, and because they have a strong community focus they are often free to the general public.

## Festivals, tourists and performers

Kirschenblatt-Gimblett (1998) writes extensively about performing culture, noting that the European tendency has been to parcel each art form separately, splitting up the senses, and ensuring that art is experienced with sustained attention and the minimum of distractions and noise. Festivals, however, offer a whole host of sensory experiences and the performance spaces are not hermetically sealed. It is a form of ethnographic, environmental performance. Festivals are described as the perfect entrée for the tourist seeking to engage with the destination and to penetrate the quotidian.

Festivals have been a cultural phenomenon for hundreds of years, dating from when a festival was traditionally a time for celebration and relaxation from the rigours of everyday existence. Traditionally, festivals were first and foremost religious celebrations involving ritualistic activities. For example, in Ancient Greece, festivals afforded an opportunity to worship deities, and prayers were offered for a good harvest or success in battle. In late-mediaeval times in Europe, festivals took on a more secular identity, and adopted a growing tendency to celebrate the greatness of men and their artistic achievements. Often, festivals would serve as a means of reaffirming or reviving a local culture or tradition, and would offer communities the chance to celebrate their cultural identity. Festivals also aim to support and promote local artists and to offer a concentrated period of high-quality artistic activity.

The idea of combining festivals with tourism dates back well over a hundred years to 1859 with the Handel Centenary Festival held in London's Crystal Palace. Adams (1986) noted that festivals and tourism have had a long history of mutual benefit. Kirschenblatt-Gimblett (1998) describes how festivals of all kinds have proliferated since the growth of mass tourism in the post-war period with the explicit intention of encouraging tourism. Rolfe (1992) demonstrated that over 50 per cent of existing arts festivals in the UK originated during the 1980s, and that this growth was at least partly aimed at increasing tourism in many tourist cities. Today, although many festivals aim to cater primarily for the local community, they succeed nevertheless in attracting tourists, and around 56 per cent of all festivals are created with a tourist audience in mind. Festivals clearly have a higher concentration of visitors in areas of the country that are already established tourist destinations, and the majority of festival organisers therefore design the programme content with the attraction of tourists in mind. Zeppel and Hall (1992: 69) state that:

> Festivals, carnivals and community fairs add vitality and enhance the tourist appeal of a destination. Festivals are held to celebrate dance, drama, comedy, film and music, the arts, crafts, ethnic and indigenous cultural heritage, religious traditions, historically significant occasions, sporting events, food and wine, seasonal rites and agricultural products. Visitors primarily participate in festivals because of a special interest in the product, event, heritage or tradition being celebrated.

Festivals can become the quintessence of a region and its people. There are, of course, problems of authenticity, compromising of artistic integrity, or trivialisation of culture: 'The more ethnographic festivals and museum exhibitions succeed in their visual appeal and spectacular effect, the more they re-classify what they present as art and risk appealing to prurient interest' (Kirschenblatt-Gimblett, 1998: 73). Getz (1994) points out that authenticity means something different for traditional as opposed to created events. In

**Plate 8.1** Bustling Indian street scene on the day of the Teej Festival, Jaipur

Source: Author

the case of traditional events, authenticity must belong to the community presenting the event. Repetition of the performance or event to meet the demands of audiences does not necessarily imply that the value of the event is being compromised. This depends very much on the perceptions and expectations of the performers and their audiences. In his comprehensive and fascinating study of performance, Carlson (1996: 195) suggests that 'performance implies not just doing or even re-doing, but a self-consciousness about doing and re-doing, on the part of both performers and spectators'. He goes on to say that 'Performers and audience alike accept that a primary function of this activity is precisely cultural and social metacommentary, the exploration of self and other, of the world as experienced, and of alternative possibilities' (p.196). Hence performances which form part of the cultural tourism experience do not necessarily imply the inevitable exploitation or objectification of performers by audiences. The relationship can be based more on mutual understanding and engagement.

Carlson (1996) also provides an interesting analysis of the significance of performance for traditionally marginalised groups, such as women, homosexuals and ethnic minorities. Tensions between self and society might be explored, including issues relating to objectification, exclusion and identity. In some ways, the development of larger audiences for local, small-scale, traditional or minority performances and events is helping to raise the profile of such groups. Carnivals are a good example, as are jazz festivals which are now universally popular. Many jazz festivals now attract large numbers of international tourists, despite jazz music traditionally being a marginalised form of music belonging to poor, black communities.

## The development of ethnic arts tourism

Lippard (1990) discusses some of the problems of nomenclature when referring to the art of 'people of colour'. The term 'ethnic' is frequently used, especially where 'black' has been deemed to be a reductionist term which fails to take into consideration the diversity of different cultures. Sarup (1996: 178) notes the polysemic nature of ethnicity, which has many meanings including the 'shared, cultural, historical features of a group', and Friedman (1999: 253) states that:

> Culture can move anywhere and be handed down to anyone, but ethnicity is about social boundaries, not about the content of what is on either side of them, not about what can be transferred from one person, or region to another, but to the way it is identified in relation to the group.

To a certain extent, it could be argued that the debates which are central to cultural politics, and the significance of race and ethnicity have been largely under-researched in the field of cultural tourism studies. Much of the research has tended to focus on post-colonial, developing countries, where tourism is sometimes viewed as a new form of imperialism, and the relationship between hosts and guests requires sensitive management (as discussed in Chapter 3). In countries where indigenous community culture and heritage form an integral part of the tourism product (e.g. Australia and New Zealand), much of the cultural tourism research has been focused on these communities, as discussed in more detail in Chapter 7. Cultural tourism has not always been viewed as a positive development in such contexts, especially with regard to Aboriginal arts. For example, Fourmile (1994: 75) states that

> Much Aboriginal artistic endeavour is commercially driven, aimed at the tourist dollar . . . with an enormous current emphasis on Aboriginal cultural tourism, one

> senses that much of the revival and maintenance of aspects of Aboriginal traditions will be dependent on, and therefore modified to suit the needs of tourism.

She also notes that tourists tend to be selective in their interest in Aboriginal arts, preferring dance groups or crafts to music, drama or literature.

South Africa is an interesting example of a country that has a rich ethnic heritage and diverse indigenous art forms. Although many of these were suppressed or marginalised during the colonial period, what is now emerging in the new post-apartheid South Africa is an exciting, hybridised arts scene, fusing elements of both indigenous and colonial culture (see Box 8.1).

In post-imperial countries, it could be argued that the emphasis in cultural tourism research has often been placed on the significance of national and regional rather than ethnic minority cultures. There are, of course, exceptions to this. For example, the work of Keith Hollinshead is particularly pertinent with regard to these issues, and warrants further reading. Despite the increasingly globalised, hybridised, multicultural and multi-ethnic nature of society in many European countries, Eurocentric or ethnocentric approaches to the management of cultural tourism may still prevail. Demographic shifts have clearly given rise to new arts audiences in large European cities, and the emergence of a global culture, aided by mass media and technology has intensified the communication between cultures, peoples and nations throughout the world. However, there is also evidence of increasing fragmentation into nationalistic movements and ethnocentrism (as discussed in Chapter 4). Many factors have influenced the construction of identity and the politics of representation in Europe, such as decolonisation, mass migration, 'Europeanisation' and globalisation. Unfortunately, in many European countries, policies of assimilation assert

## Box 8.1 Grahamstown National Arts Festival

The annual Grahamstown National Arts Festival on the East Cape of South Africa is a diverse, multicultural festival which involves a whole range of art forms, including theatre, dance, opera, music, poetry, fine arts and crafts. It was established in its current format in 1974, and at that time included about sixty events. There are now around 600 with close to 1800 performances, and the number of supporters has grown from several hundred to around 100,000.

The city of Grahamstown was founded in 1812 by Colonel John Graham as a military headquarters for British troops. It is a typical colonial-style English nineteenth-century cathedral town, but is located in a setting that is unmistakably African. Unlike many other South African cities, the poor black townships are situated close to the smart suburbs and tourist attractions, hence providing a reminder of the deep class and racial divides that have characterised the country for so long.

In the early nineteenth century, the colonial rulers established a centre for conferences and festivals. It placed emphasis on Western art and local colonial history to reflect the South Africa of its time. However, over the years, the festival has lost its Eurocentric focus, and by the 1980s it had become something of a forum for political and protest theatre. Although the event has become somewhat depoliticised in more recent post-apartheid years, it now reflects the South African culture of today, which is a rich tapestry of indigenous African, Western and Asian cultures. This includes ballet and Indian dancing, township jazz and organ recitals, Zulu beadwork and watercolours, African operas, and other hybridised music and dance forms, often of an experimental nature.

Source: Adapted from Standard Bank National Arts Festival (2001)

that people from different cultures and races should adjust to the dominant national or European culture, as stated by Lavrijsen (1993: 16):

> Europe has always known historical, cultural and experiential differences within and between communities, nations, regions and cities. The idea of one European homogenised culture is a myth, as is the idea of homogenised cultural identities within specific non-European ethnic groups. However, because of the power structures within the art world, and its increasing orientation to the mass media and to commercial and market demands, there is a tendency towards homogeneity in national, European and global cultures.

More often than not, the political and social status of immigrant and ethnic groups is uncertain, and their power to represent themselves and their interests is limited. Complex relationships of dominance and suppression tend to govern the extent to which such groups are granted political or cultural autonomy and support. Priorities in policy-making and funding tend to favour initiatives and projects that represent national interests, and usually those of white, middle-class society. For example, it is clear that many national organisations such as the Arts Council of England have struggled to accept a more inclusive and democratic definition of minority and ethnic arts and culture. Such groups may consequently feel under-represented, misrepresented or stereotyped, and hence unable to develop a clear sense of cultural identity.

Within contemporary British cultural studies, identity construction and representation have become recurrent themes, with racial and ethnic issues gaining significance from the mid-1980s onwards. Black writers and theorists, such as Hall, Gilroy and Mercer have focused extensively on such issues in their work. In her seminal report, Khan (1976) emphasised the importance of the ethnic arts and cultural activities that Britain tends to ignore. Owusu (1986) took this work further, focusing particularly on black communities. Hewison (1997) discusses the various ways in which 'subcultures' (e.g. ethnic, women, youth and gay groups) challenged the prevailing cultural hegemony in the 1970s in Britain. More recent reports have accorded such issues greater priority. For example, Jermyn and Desai (2000) discuss how barriers to ethnic minority participation in the arts could be removed, and the Arts Council of Great Britain has devoted considerable attention to such issues in its recent policy documents.

These debates are not, of course, confined to the UK and Europe, although the nature of cultural politics in these countries is influenced largely by their imperial pasts. It is recognised that there are a large number of international cities that have become more and more ethnically diverse, with second and third generation immigrants playing a key role in all aspects of cultural and arts development. This is particularly true of large cities in the USA, as well as Canada, Australia and New Zealand. However, it is also worth noting the increasingly globalised and culturally diverse nature of many Asian and South American cities (e.g. Tokyo, Singapore, Rio de Janeiro) (Box 8.2).

One of the key challenges within the arts and cultural industries is clearly the need to create activities, exhibitions and collections with which the maximum number of a country's culturally and ethnically diverse population can identify and participate in. This includes the appropriate interpretation and representation of indigenous and ethnic collections in museums and galleries; the provision of space and funding for ethnic and minority performing artists; and the support of local festivals and special events. The development of cultural tourism can variously support or compromise the nature of many festivals or events depending on how it is managed. The following sections will discuss in more detail some of the issues relating to festival and event tourism, especially those which are organised by ethnic or minority groups.

## Box 8.2 WOMAD: a global arts festival

WOMAD stands for World of Music, Arts and Dance, and is a festival which aims to bring together and to celebrate the music, arts and dance of a diverse range of countries and cultures throughout the world. The idea was inspired originally by the musician Peter Gabriel, who set up the first WOMAD festival in 1982. He stated that the aim of the festivals was to introduce an international audience to a number of new and established talented artists, and to celebrate the diversity of the globe:

> The festivals have allowed many different audiences to gain an insight into cultures other than their own through the enjoyment of music. Music is a universal language, it draws people together and proves, as well as anything, the stupidity of racism.

Since the first WOMAD festival in the UK in 1982, WOMAD established a unique nomadic identity, and began to be presented internationally in 1988 in Denmark and Canada. Since then, WOMAD has staged more than ninety events in twenty different countries and islands. These include Australia, North America, South Africa, Japan and several European countries.

The festivals often last all weekend and tend to be family-orientated, active and diverse musical events. The festival features simultaneous performances of music and dance on two or three stages, many of which are hosted by visiting artists. There is a 'global village' of food and merchandise stands. Participatory arts workshops and special events for children are organised, many of which are educational, aiming to raise awareness of and to create an interest in the exciting potential of a multicultural society.

WOMAD is arguably the premier worldwide, multicultural music festival. It not only brings diverse musicians together, providing them with new audiences, but it also aims to broaden the minds of its audiences. As stated by Peter Gabriel:

> WOMAD is about discovery, including exposure to the culture and foods of different places. In the same way they used to say 'Travel broadens the mind', I think WOMAD can broaden the soul if you're open.

Source: Adapted from: WOMAD (2001)

## Ethnic festivals, communities and regeneration

The development of ethnic festivals can sometimes help to raise the profile of local community groups, leading to a greater understanding of and interest in their culture. They can also be highly attractive for tourists, as well as acting as catalysts for urban regeneration. Bradford Mela is a good case in point (Box 8.3).

Bradford is clearly a city that is keen to regenerate itself using culture and cultural events as part of the process. Chapter 9 will discuss such developments in more depth, but it is worth noting here that festivals and cultural events can play an important role in the urban regeneration process. Past regeneration strategies have generally tended to marginalise or bypass local community needs, but there has been a recent shift towards a more community-focused approach to regeneration.

In terms of socio-cultural impacts, festivals can play a key role in local community development. Table 8.2 summarises some of the costs and benefits.

The concept of identity is sometimes a key issue for younger community members, especially children of first or second generation immigrants who often have a sense of dual

## Box 8.3 Asian arts and the Bradford Mela

Bradford is an example of a northern industrial English city which has used its diverse culture and ethnic arts to help regenerate the city and to develop tourism. Hope and Klemm (2001) describe how Bradford developed its tourism industry almost from scratch, and has consequently never been a 'natural' tourist attraction. Until 1980 Bradford did not have a tourist industry at all; however, it was recognised by the city council that it could appeal potentially to the short break market. Consequently, Bradford started to promote its industrial heritage, coupled with the culture of its large Asian community. Bradford now markets itself as 'Vibrant, diverse and full of surprises'. Much of its unique tourism product derives from the presence of its multi-ethnic population and tributes to their rich heritage and culture. One of the biggest selling points has been Asian cuisine, and a booklet called *Flavours of Asia* was produced. This detailed fifty or so Asian restaurants, several curry tours, the largest Asian store in Europe, some sari centres, a brief history of various Asian religions and the patterns of immigration to Britain (Urry, 1990). Bradford is now known as the curry capital of England! Indian film and music is so prevalent that one cinema specialises in Bollywood and two others, including the National Museum of Film, Photography and Television, stages the Bite the Mango festival with pictures from Black and Asian countries. In addition, in April 2001, the Mughal Water Gardens were officially opened. These are designed in the architectural style of the Mughal Dynasty which ruled northern India from 1526 to 1858.

However, the best-known Asian attraction in Bradford is the Mela Festival which is now one of the biggest Asian festivals in Europe. It attracts visitors from all over the country, and is viewed by many as the Asian equivalent of the Notting Hill Carnival in terms of cultural value. The first Bradford 'Mela of Art and Culture' took place in September 1988. This was a one-day free event which included Asian music and dance, fashion, poetry, theatre, crafts, food stalls and a Punjabi-style fairground. An Educational Workshop programme was also organised, and this involved thousands of schoolchildren. The event attracted around 10,000 people. In 1989 the Mela became a two-day festival with a finale event which included the Lord Mayor's Carnival Parade. There were two stages and 100 stalls, and 60,000 people attended over the weekend. The festival concluded with spectacular fireworks.

In 1992 the dates of the Mela changed to June/July, and it then attracted an audience of around 120,000. There were over eighty performances, with a large number of international artists, including troupes from India, Pakistan and the East. The diversity of the programme has also been one of its most significant features, and there were food stalls and market areas to match. By 1997 visitor numbers were up to 150,000.

Bradford is currently bidding for European Cultural Capital status for 2008, and it is selling itself partly on its multicultural, multi-ethnic heritage. The successful Mela is, of course, one of its main selling points.

Sources: Adapted from http://www.bradford.gov.uk/ http://www.oriental.legend.org.uk/mela.html

identity. Festivals can be used to increase racial tolerance through cross-cultural exchange and education. The involvement of young children through local schools in various workshops, events and processions can help to foster an understanding of cultural diversity and community celebrations.

Raising awareness both of similarities and differences is a major challenge for those working in community and cultural education. It is important for different community groups to feel comfortable with the identity of others, and to accept and tolerate difference. One of the most interesting and accessible ways of celebrating both similarities and differences is through the grouping of community festivals which have a common theme

**Table 8.2** ***Costs and benefits of community festivals***

| *Social benefits of community festivals* | *Social costs of community festivals* |
|---|---|
| • Showcase for new ideas<br>• Opportunity to learn new things<br>• Enhancement of community image<br>• Increased community pride<br>• Strengthening of community identity and cohesion<br>• Working together and sharing ideas among community groups<br>• Improved quality of life | • Intrusion into and disruption of community life<br>• Increased noise, traffic, pollution<br>• Increased crime and vandalism<br>• Strain on limited financial resources<br>• Inter-community rivalry<br>• Some groups benefit more than others<br>• Highlighting of negative cultural stereotypes |

Source: Adapted from Delamere *et al.* (2001)

(for example, festivals of light), and which are celebrated by diverse communities in different ways.

Representation is clearly a key issue when planning ethnic festivals. Representation of the diverse cultural activities and art forms of different community groups is an important consideration, but it is perhaps the quality of the events that is most significant when programming festivals. It is probably true to say that the majority of local festivals tend to be under-funded; therefore the profile and quality of performers cannot always be guaranteed. However, many local festivals are now aspiring to include some high-profile, international artists in their programmes, if only to enhance the image of the festival and to attract sponsorship. The exposure of local people in deprived areas to high-quality cultural events is also important if aspirations towards increasing access and inclusion are to be achieved.

Earlier, the involvement of young people in community festivals was mentioned as being a priority for many festival organisers, cultural and regeneration officers. Education of younger members of the community is of paramount importance, and can help to foster better social and ethnic integration for the future. Many festivals now include workshops and education programmes for young people, which are often linked to schools. Inter-generational conflicts can also be alleviated through festival events, which involve community members of all ages, races and genders.

The involvement of women in ethnic festivals is particularly worth mentioning, as it is significant that festivals are perhaps the most family-orientated and child-friendly of any cultural or arts events. Many Asian and Muslim women are often less able to participate in cultural events outside the home than other groups; therefore festivals can afford them an ideal opportunity for participation and involvement. This might include preparation of costumes or catering, for example.

The aim of many festivals is to enhance the image of an area and to 'put it on the map'. This is especially true of deprived areas of inner cities where local residents may not feel very positive about their neighbourhood. The other priority in social terms is to improve the quality of life for local residents. It is clearly important to raise the profile of an area and enhance its external image if business investment, sponsorship or tourism are to be attracted. However, the internal image of an area should be given equal consideration if people are to feel positive about their neighbourhood, their culture and their identity. In a general context, this might include the provision of adequate cultural and leisure facilities, education and skills training programmes, improvements in safety and security, or boosting the evening economy. Festivals can perhaps act as catalysts for all of these developments.

## The development of carnivals as cultural tourism events

Bakhtin (1965) is the one of the main theorists associated with the concept of carnivals. He describes carnivals as being 'Rabelaisian' in nature, whereby participants and spectators are released from the constraints of everyday life to engage in sensuous, hedonistic and licentious pleasure. Sampson (1986: 34) writes of the typical Trinidadian Caribbean carnival as being:

> An aphrodisiac – a gigantic, erotic road and stage show – a breezing, reeking melodramatic atavism. It is high and low musical drama, peppered by mild emotions, lascivious designs and gushing alcohol. It's the Alpha and Omega of mass hysteria: unrestrained jollity; carefree carnality; and calculated (sometimes spontaneous) crazy behaviour. It's an upsurge where for two fleeting days the players can enjoy in fantasy what is denied to them in reality.

Miles (1997) describes the 'polyphonic' nature of carnivals and festivals, suggesting that it is the ideal forum for representing a multiplicity of perspectives and the expression of ethnicity. Alleyne-Dettmers (1996) traces the history of carnival, which is often based on historically and culturally specific models which are born out of the oppressive context of European imperialism, colonisation and slavery. This is particularly true of Caribbean carnival, which is now a particularly vibrant event throughout the region. The biggest carnival in the world in Rio de Janeiro combined the cultures of the colonial Portuguese rulers with that of African slaves and native Indians, resulting in a rich, colourful, hybridised form of Afro-Brazilian culture which is expressed through one of the biggest and best-known cultural events in the world. Large numbers of tourists are keen to experience the carnival of these countries, to enjoy the spectacle, and to free themselves from the constraints of their everyday lives.

European carnivals are perhaps more sedate, less frenzied affairs; however, they are equally popular with both residents and tourists alike. In Europe, the history of carnival can be traced back to the concept of Carnivale in Italy and the masquerade balls of the sixteenth century, where the masked Mardi Gras derived partly from the 'Commedia dell'Arte'. Nowhere in Italy was this phenomenon as spectacular as in Venice (Box 8.4).

### Box 8.4 Venice Carnival

The Venice Carnival was always one of the most popular feasts in the city, and records documenting this event go back as far as the eleventh century. The Carnival was always lavish, and by the eighteenth century it was known all over Europe. It was characterised by its magnificent masquerades, masked balls and feasts, rather than by street processions and parades which were common elsewhere. This was due largely to its situation as a town built on water and connected by bridges. However, the whole town was transformed into a vast theatre, full of music, dance and festivities. The various 'campi' or small squares throughout the town were traditionally used to stage various events, as they still are today. However, in 1769 when the Republic fell, the vitality of Venice was temporarily lost, and the tradition of Carnival abandoned.

In 1980 the Venice Carnival was revived as a community and tourist festival, partly in order to address problems of seasonality, but also to revive a vibrant tradition. During the Carnival period Venice Art Biennale organises a special programme of cultural events while the La Fenice and Goldoni theatres offer a full and varied programme of opera and drama. Free

continued

**Plate 8.2** Venice Carnival: age-old tradition revived for the sake of tourism

Source: Author

open-air events such as music concerts, street theatre and dance are also staged in various locations. Roiter (1991: 43) states that:

> Carnival has rediscovered its ideal setting in Venice; the masks have returned to identify themselves with the fabric and the authentic spirit of the city; there has been an effective explosion of revelry, retaining the primitive features of the feast of transgression and responding to the popular need for a pause, a brief respite from the demands of everyday living.

One of the most positive benefits of the Carnival has been the revival of the skilled craft of traditional mask-making, which is now a very prominent feature of Venice's retail and souvenir industry.

However, the revival of carnival is not without its problems. Crowhurst Lennard and Lennard (1987: 76) describe how:

> As a recent tradition the rebirth of Carnival – especially its unanticipated success – has given rise to similar controversies in Venice as in other cities where festivals have either been revived or sometimes even newly invented, that is, questions about authenticity, commercialisation, and a balance between resident and tourist involvement and needs.

It is evident that the February Carnival has done much to address the seasonal nature of Venice's tourism industry, although some might argue that Venice did not need to attract more tourists to its already fragile and over-visited environment. The local residents' relationship to the Carnival perhaps warrants further research, since it is uncertain as to how far this revived event is as popular with the locals as it is with tourists. Local residents are now given little respite from tourism, and, for all its beauty, Venice is arguably becoming more of a tourist museum than a thriving, residential town.

Kirschenblatt-Gimblett (1998: 77) suggests that tourism can sanitise the carnival experience by aestheticising it and treating it ahistorically: 'Carnival represented is carnival tamed.' There have recently been concerns about the Notting Hill Carnival in London, which has become more and more popular with Londoners and tourists alike, now attracting around two million tourists (Box 8.5). Some have argued that the event has outgrown the local area and should be moved to central London. Unfortunately, the event has recently sparked violence, including numerous arrests and even murders. However, the event clearly has its roots in what was once a predominantly Afro-Caribbean area of London, and removing it would be something of a wrench for local communities, as it would displace a cultural event with clear links to the area's heritage. Notting Hill has already been the focus of a controversial gentrification process which was exacerbated by the success of the movie *Notting Hill*. There appears to be some discontent and disillusionment among the local Afro-Caribbean population who feel (perhaps rightly) that the area is being taken over increasingly by white, middle-class communities, and this includes the Carnival.

## Box 8.5 Notting Hill Carnival

The Notting Hill Carnival is now the second largest carnival in the world after Rio, attracting over two million visitors. It has its origins in the culture and heritage of the local Afro-Caribbean communities, particularly those from Trinidad. The Carnival was traditionally developed as a means of expressing and celebrating the vibrancy and diversity of Caribbean culture. The concept of carnival has also been highly politicised and linked to the protection of local heritage and identity. As stated by Owusu and Ross (1987: 39), 'it is the celebration of emergence, an affirmation of survival and continuity'. Khan (1976: 7) writes of how the carnival appealed both to Caribbeans and indigenous British alike:

> Its virtue is that it not only revitalises life, but that it also provides a bridge for the indigenous British. For a couple of days it is possible to participate in a 'jump-up', to share in an experience that is no common and everyday English thing, and – possibly – to catch some understanding of the enjoyments and values that gave rise to that form.

The Carnival has not been without its problems, particularly in the mid-1970s when objections to its growth were lodged by the resident white community. Race relations were fraught, especially as rioting and unrest had led to a need to increase security and police presence. It was suggested that the Carnival had outgrown the local area and might have to be moved, or, worse still, discontinued, which the Afro-Caribbean community naturally campaigned against. As emphasised by Owusu (1986: 15):

continued

> the Notting Hill Carnival and other offshoots across the country stand as significant symbols, crucial antitheses to this history of state repression. Their survival and development are therefore not the concern of the Black Community alone, but of all who appreciate and cherish the vitality of popular creativity.

The Notting Hill Carnival has consequently survived, but continues to be a source of contention at times. Unfortunately, the event has traditionally been under-funded as are many such ethnic events, and its future survival has frequently been threatened until the last minute by the apparent unavailability of sponsorship. Although the event has become increasingly popular with large numbers of tourists, they do not make any financial contribution to the staging of the event. Some have also argued that the presence of tourists, especially so many white tourists, has compromised the cultural significance and atmosphere of the Carnival. For example, Errol (1986: 17) cites a local taxi driver who laments 'too many goddam tourists who can neither chip nor jump comin' here spoilin' the carnival'.

However, despite the ongoing problems with the growth of the Carnival, increasing visitor numbers, inadequate funding, and recent violence, its future seems to be safe in the hands of London's Mayor, Ken Livingstone, who has always been a vociferous supporter of ethnic and minority events. In 2000, he stated that:

> The Notting Hill Carnival has become London's largest cultural festival and a showpiece for the vitality and diversity of the capital. It is a testament to the international character of London as a truly world city that our most popular street festival is inspired by the traditions of the African Caribbean community and enjoyed by Londoners and tourists of every conceivable ethnic origin.
>
> (Livingstone, 2001)

Livingstone pledged to increase access to financial support and to strengthen security arrangements without compromising or curtailing the vibrant and spontaneous nature of the event. The spirit of Notting Hill Carnival lives on – at least for now.

It is clear that ethnic events can help to reinforce a sense of cultural identity and community cohesion, but that the over-commercialisation of such events (for example, for the purposes of tourism) can somehow detract from the original purpose and meaning of the event. Hence communities must be given control of the organisation and development of such events if they are to retain their local and ethnic significance.

The final section of this chapter will take a brief look at the growth in popularity of gay arts events, many of which are becoming more mainstream and tourism-orientated, but which also face the threat of misappropriation.

## The growing popularity of gay arts events

The development of gay cultural tourism has been a relatively under-researched phenomenon, although it is recognised that, in terms of demand and motivation, the gay market is potentially as diverse as any other. However, a number of gay events are becoming increasingly popular both with the gay and mainstream markets. Chapter 9 will analyse briefly the way in which a number of gay spaces have been created in cities, often as part of urban regeneration schemes. Many of these are centred on the strength of the 'pink pound', and can therefore be some of the most gentrified areas in a town. A good example of this is Manchester's Gay Village, which has become the main focus of the city's vibrant nightlife. Indeed, a number of cities both in Europe and internationally can claim to have a gay quarter which has a thriving evening economy, not to mention daytime leisure and retail provision. This is certainly true of London, Madrid, Amsterdam, San Francisco, Sydney, Rio, to name but a few. Many English seaside towns are also actively promoting gay tourism as a means of establishing a new image or identity. Examples include Brighton, Blackpool and Bournemouth, which have all been promoted as gay destinations by the British Tourist Authority.

The increasing popularity of gay areas, especially with women and heterosexual couples, has sometimes led to concerns about the 'de-gaying' of certain spaces, in much the same way that Notting Hill Afro-Caribbean residents are concerned about the increasing white, middle-class presence. This is certainly a contentious issue, but it is perhaps symptomatic of the growing popularity of gay culture and events. The most significant examples of gay cultural events that draw a mainstream and tourist audience, as well as a gay one, are the Gay Pride or Mardi Gras events. Originally, such events were highly politicised, and the marches which accompany such events still tend to be so. These were largely a forum for the gay community to present a united front, and to express and assert their identity and rights publicly. However, in most cases, the lively celebrations that follow are becoming more of a public party than anything else. Indeed, the London Gay Pride event was criticised by some in 2001 for being too depoliticised and commercial, especially as it attracted such a large number of heterosexual revellers. The Gay Mardi Gras concept is quite well developed in Britain, and usually involves a day of concerts, dance tents, funfairs and food stalls. London is still the biggest event, and it has consequently become a ticketed rather than free event. In other cities where the event is relatively new, such as Birmingham, the events are still free. However, the concept is not limited to the UK, as the case study outlined in Box 8.6 shows.

## Box 8.6 Gay Mardi Gras in Sydney

Like most Gay Mardi Gras events, the Sydney Mardi Gras had its origins in a march that took place in 1978 to mark Gay Solidarity Day. The following year the Mardi Gras name was adopted, and the idea of this event as a celebration was mooted in 1980. The emphasis was on celebration, fun, education and politics. In March 1981, the Mardi Gras Parade was held before a crowd of 5000. By 1982, this number had risen to 10,000, and the first post-parade party was held, attended by 4000 people. Support for the event was growing all the time, although the straight media's fear of AIDS in the mid-1980s put pressure on the gay community to cancel the event, which they declined to do. These threats notwithstanding, the event has gone from strength to strength. By 1987, the festival consisted of thirty-five separate events, and the estimated parade crowd was 100,000. During 1988, the Bicentennial year in Australia, an Aboriginal float led the Mardi Gras parade with an Aboriginal man dressed as Captain Cook. By 1989, the Festival ran for two weeks and the estimated parade crowd was 200,000; by 1991, it had become the largest parade in Australia's history. A National Gay and Lesbian Film Festival was held for the first time as part of the Arts Festival. In 1992, the parade crowd was up to 400,000, and the Festival had become the largest lesbian and gay festival in the world, running for four weeks. By 1993, the Parade had become the largest night-time outdoor parade in the world, and it was estimated in a Mardi Gras Economic Impact Study that the total impact for the Australian economy was around $38 million. Throughout the 1990s, the event received increased publicity, being filmed for mainstream

continued

television in 1994, and launching a hugely popular Mardi Gras Party Anthems CD in 1995. In 1998, Mardi Gras celebrated its twentieth anniversary, and a comprehensive economic impact study estimated a total impact of $99 million on the city. Not only had the pink pound contributed enormously to the city's cultural economy, but the event had secured mainstream financial, political and moral support.

Source: Adapted from Sydney Gay Mardi Gras (2001)

## Conclusion

This chapter has demonstrated the important role festivals and special events play in the development of cultural tourism. Festivals and events are often more accessible to the masses than other art forms, as they provide an open forum for the celebration of life and the continuity of living. In many cases they can also be an expression of local community culture, traditions and identity. Although care must be taken by the community to ensure that the authenticity and enjoyment of their celebration is not compromised by tourism, it is clear that new audiences can often be created for ethnic and minority cultural events. As many of these events may previously have been threatened or marginalised politically or financially, tourism can help to raise their profile and encourage support. Although such events tend to be free for locals and tourists alike, hence encouraging access and participation, they can also make a significant contribution to the local economy. As discussed in the early part of this chapter, the relationship between the arts and tourism is not always a harmonious one; however, it is interesting to note that in the case of certain art forms, such as minority and ethnic events, tourism can make a positive contribution to cultural continuity.

# 9 Cultural tourism and urban regeneration

> Culture rising up from below, not passed down from above, is . . . an important characteristic of the new cultural planning impetus, the new movement for community enrichment. This assures popularity, pluralism, and whatever the antonym for elitist is.
>
> (Von Eckardt, 1980: 140)

## Introduction

The aim of this chapter is to analyse the role that cultural tourism can play in the regeneration of cities, especially those that have suffered as a result of de-industrialisation. Over the past few decades there has been a significant shift from a manufacturing to a services-based economy in many major cities, particularly in the UK, other parts of Western Europe and North America. This has led to the growing importance of tourism, leisure and cultural activities as sources of revenue and catalysts for socio-economic development.

Cultural tourism can provide alternative sources of revenue for cities where traditional industries have declined, and whose economies, environment and communities have suffered as a result. The use of cultural events and 'flagship' initiatives is becoming a common means of transforming cities, attracting inward investment and enhancing image. A thriving cultural economy can often improve the socio-economic status of a city and contribute positively to local community life.

The aim of this chapter is to analyse some of these issues in more detail, focusing in particular on former industrial cities in Western Europe and North America where the concept of regeneration has been researched in some depth.

## The role of cultural regeneration in urban and economic development

The phenomenon of cultural regeneration in the UK, North America and Western Europe from the late 1970s onwards has been recognised by a number of writers and researchers (see e.g. Zukin, 1988; Wynne, 1992; Bianchini and Parkinson, 1993; Landry and Bianchini, 1995). Much of the research has tended to focus on the way in which the so-called 'cultural industries' contribute to the economic development of cities, especially de-industrialised cities, which are moving towards a more services-based economy. The role culture and tourism can play in the physical, economic, social and symbolic transformation of urban spaces is significant. Researchers have tended to focus on the development of cultural regeneration strategies for cities in decline, which have used culture to expand their economic base, enhance their image, and contribute to social integration and community cohesion.

The cultural regeneration process usually forms part of broader urban regeneration strategies, and may even be central to these. This is especially true of cities that are keen to develop and promote cultural tourism. The effects of cultural regeneration initiatives are often cumulative, encouraging inward investment in improved infrastructure, environmental developments, and better local facilities and services. Over the past few decades, urban and cultural regeneration strategies have become increasingly integrated, as indicated by Worpole (1991: 143): 'Urban policy in the late twentieth century is now inseparable from cultural policy', and McGuigan (1996: 2): 'Cultural policy [cannot] be treated satisfactorily in isolation from the wider economic and political determinations operating upon culture and society'.

Fox-Przeworski *et al.* (1991: 250) suggest that 'Clearly, there is no such thing as a single set of measures that can bring about successful urban economic regeneration for all cities.' A 'cultural planning' approach, which is being increasingly adopted, aims to co-ordinate cultural policy with other urban policies (e.g. economic, environmental, social, political, educational, symbolic) to ensure a more integrated development (Bianchini, 1993). Evans (2001) provides a fascinating and comprehensive analysis of the development of arts and cultural planning within the context of urban renaissance. Von Eckhardt (1980: 142) emphasises the integral nature of cultural planning: 'Effective cultural planning, in short, involves all the arts . . . the art of architecture, the art of urban design, the art of winning community support, the art of transportation planning, and the art of mastering the dynamics of economic development.'

Harvey (1989) argues that cultural regeneration is not a panacea for urban decline following de-industrialisation. However, although cultural and economic policy are not synonymous, they are inextricably linked in the context of urban development. Cultural planning has a significant economic dimension, as stated by Von Eckhardt (1980): 'Good cultural planning, like good city planning must be good economic planning.' Smith (1996: 57) suggests that 'the so-called "urban renaissance" has been stimulated more by economic than cultural forces'. However, in the past, too much emphasis has perhaps been placed on the economic imperative in regeneration strategies. Fox-Przeworski *et al.* (1991: 237) claim that 'the most obvious foundation for successful local economic regeneration is an objective appraisal of the problems and opportunities confronting the local economy'. As a consequence, social, cultural and welfare issues which are so important to local communities have often been overlooked, as the following section will demonstrate.

## The role of local communities in cultural regeneration initiatives

Many authors have argued that local communities often fail to benefit from urban and cultural regeneration projects (e.g. Zukin, 1988; Bianchini, 1990; Colenutt, 1991; Keith and Rogers, 1991; Castells, 1994; McGuigan, 1996). The DETR (1997) refers to the local community in the context of urban regeneration as those people who are living or working within target areas. If regeneration initiatives are to succeed ultimately, then local communities should clearly benefit. For example, much of the political rhetoric in the UK in the 1980s claimed to prioritise community needs; however, these promises were not necessarily borne out in reality. Much of the regeneration of UK cities at that time was managed through Urban Development Corporations (UDCs). Following the problematic regeneration of London's Docklands, a House of Commons Employment Committee (1988) report concluded that 'UDCs cannot be regarded as a success if buildings and land are regenerated but the local community are by-passed and do not benefit from regeneration' (par. 89). The Community Development Foundation (1995: 11) later echoed this, stating:

'Measures to improve the competitiveness of the local economy must take account of the existing economic, social and environmental context. Where jobs are to be created or housing improved, it should be clear that the existing population will be the main beneficiaries.' Bianchini (1993: 212) also stressed the importance of local community interests in urban regeneration initiatives in Western Europe: 'an explicit commitment to revitalise the cultural, social and political life of local residents should precede and sustain the formulation of physical and economic regeneration strategies.'

Unfortunately, it is recognised generally that local residents often fail to benefit directly from urban and cultural regeneration projects. In some cases, high-profile regeneration projects can even exacerbate social and cultural polarisation within cities, creating what Castells (1994) referred to as a 'dual city'. In other cases, regeneration initiatives simply 'shift' social problems from one part of a city to another: 'Displacement removes social problems and rearranges rather than ameliorates the causes of poverty, environmental decay and the loss of neighbourhood vitality – problems are moved rather than solved' (Atkinson, 2000: 163).

Robins (1993: 321) refers to the development of 'art, culture, consumption and a cappuccino lifestyle', which benefits only an elite group. Harvey (1989) criticises prestige arts-led regeneration projects that mask social and welfare problems in deprived areas, and Bianchini (1990: 238) states that 'In the British context, arts-led regeneration initiatives coexist with the erosion of welfare benefits and the growth of spatially segregated "underclasses"'. Zukin (1988) refers to 'quixotic' urban renewal projects in cities such as Detroit, which failed to create sufficient economic benefits for local communities. McGuigan (1996: 99) comments on the notion of 'civic boosterism', and the emphasis that is placed on the needs of professional and managerial classes and tourists, rather than on the socially excluded urban underclasses:

> such urban regeneration, in effect, articulates the interests and tastes of the postmodern professional and managerial class without solving the problems of a diminishing production base, growing disparities of wealth and opportunity, and the multiple forms of social exclusion.

Over the past two decades many urban and cultural regeneration strategies have placed emphasis on private sector initiatives and investment, especially where residential developments have taken place. The consequent 'gentrification' of localities and public space has often led to the displacement and marginalisation of local communities. McGuigan (1996: 104) criticises the way in which the Thatcher government in the 1980s and the Urban Development Corporations (UDCs), especially in the East End of London, encouraged the widespread privatisation of public space: 'Such privatisation of public space erodes urbanity and social cohesion.' Sudjic (1993) similarly described the problems of private property-led urban regeneration along the Thames in the East End of London. Zukin (1988) refers to the displacement of lower-income residents due to residential developments and rising property prices in the USA in the 1970s. Bianchini (1993: 202) refers to examples of Western European cities such as Frankfurt that have developed 'cultural districts' which '[have] generated gentrification, displaced local residents and facilities, and increased land values, rents, and the local cost of living, as measured – for example – by the prices charged by local shops'.

Burtenshaw *et al.* (1991: 39) describe the shift in European urban planning policy from the 1970s onwards, where 'almost everywhere gentrification became a highly visible component of a fundamental transformation within the city core, in turn mirroring wider changes in the reshaping of the advanced capitalist countries'. 'Gentrification' is indeed a concept for which many urban regeneration projects have been criticised in recent years.

The term *gentrification* is believed to have been first used by Ruth Glass in 1964, and was popularised in the 1970s. She described it as a process whereby 'all or most of the original working-class occupiers are displaced and the whole social character of the district is changed' (Glass, 1964: xviii). Gentrification is clearly a complex and contradictory process. As stated by Keith and Rogers (1991: 24): 'Gentrification benefits selectively, takes away with one hand as it gives with another, bestowing respectability at a cost of displacement. The inner city is transformed, but for whom?' The case study outlined in Box 9.1 provides a clear example of the gentrification phenomenon and its various ramifications.

The result of such developments may eventually lead to the erosion of the social fabric of a locality and its communities. Bauman (1998: 24) argues that increasing globalisation and the changing sense of locality and public space have engendered a fragmentation of community cohesion: 'Far from being hotbeds of communities, local populations are more like loose bunches of untied ends.' This may have serious implications for the sense of community identity. Worse still, it may incite feelings of anger and aggression whereby urban spaces can become sites of conflict and tension: 'Urban territory becomes the battlefield of continuous space war. . . . Disempowered and disregarded residents of the "fenced-off", pressed-back and relentlessly encroached-upon areas, respond with aggressive action of their own' (ibid.: 22).

Massey (1994: 163) refers to Baudrillard's visions of hyperspace and globalisation, and postmodern theories of 'power-geometry' and 'time–space compression', arguing that 'most people actually live in places like Harlesden or West Brom. Much of life for many people, even in the heart of the First World, still consists of waiting in a bus-shelter for a bus that never comes.' It is probably true that many local communities are much more likely to be concerned about better bus routes, more car-parking space, the quality of local supermarkets, affordable housing, and an increase in local green spaces and recreational areas, rather than a proliferation of gentrified leisure spaces and cultural attractions.

However, Worpole (1991: 148) suggests that 'Urban renewal is not about gentrification but about new dynamic mixes of activities and uses, based on cultural diversity. It is about cultural production.' It is worth considering how far cultural regeneration and the development of tourism, cultural and leisure activities can offer communities the chance to reclaim urban spaces, to assert their identity, and to benefit from cultural initiatives, attractions and events. Worpole (1992: 99) suggests that:

> As with economic development, tourism strategies have the potential not only to create jobs, but also to enhance the quality of life for a town's residents as well as its visitors. All tourism strategies start with the question of self-image, and that is as much an indicator of local concern to residents as to visitors.

## The changing policy context for cultural regeneration

The development of urban cultural policy clearly reflects changing priorities in the regeneration context. Bianchini (1999) traces the historical trajectory of urban cultural policies in Western Europe, demonstrating that cultural regeneration is a relatively new phenomenon.

- *The 'age of reconstruction': late 1940s to late 1960s*
  Emphasis was placed on economic growth, physical and civic reconstruction, and welfare planning following the Second World War. Urban cultural policies tended to focus on educating and 'civilising' the public through mainly 'high' or traditional culture (e.g. opera, museums, civic theatres).

## Box 9.1 Bypassing the local community in London's Docklands

From an economic perspective, the Docklands regeneration scheme was perceived by many as being akin to those that took place in other parts of the country. For example, Wood (1986) describes the experience of Docklands as being more like that of coal-mining, shipbuilding and steelworking areas elsewhere in Britain than other areas in London. In five Docklands boroughs between 1971 and 1978 there was a loss of 53,000 manufacturing jobs as well as almost 30,000 jobs in transport industries. The regeneration process transformed an area that was economically and socially in decline into an architecturally and financially successful scheme, which generated both employment and business opportunities. Hall (1989: 213) stated that 'Docklands are a triumph of planned inner-city regeneration, of a kind'.

Unfortunately, from a community perspective, Docklands is thought to have been largely unsuccessful. Urban Development Corporations (UDCs) were heavily criticised for their apparent determination to privatise public space, and their failure to provide for local communities (e.g. Colenutt, 1991; Sudjic, 1993; McGuigan, 1996). Coupland (1992: 161) describes the regeneration of Docklands as being an example of a project where local communities were largely overlooked by the LDDC (London Docklands Development Corporation), claiming that few benefits appeared to have trickled down: 'The huge glass- and marble-clad offices have little of relevance for the local community, and represent a long-term monument to how "regeneration" can become a disaster in less than a decade.'

McLaren (1992) laments the lack of open and green spaces in London, especially in Docklands where it is estimated that only 6 per cent of the land is open space compared to 11 per cent in the rest of London. Punter (1992: 82) criticises Docklands for the almost total absence of usable urban space and lack of provision of public spaces. Crilley (1993: 153) reiterates the point: 'Canary Wharf is not designed to allow for the outdoor activities and political demonstrations that are part of the history of the social use of London's public spaces.' He argued that the spaces created are largely sanitised, and mask an uglier reality of deprivation and lack of public provision.

Unfortunately, Docklands has been viewed largely as a private property-led, 'gentrifying' scheme which has mostly benefited an elite group, bypassing the needs of deprived members of local communities. It is clear that subsequent regeneration initiatives in this relatively deprived part of London have learnt from this model, and are attempting to prioritise local community needs in their developments. While it is difficult to measure the 'trickle-down' effects of regeneration, particularly the more indirect benefits, Colenutt (1991: 34) states that in the case of Docklands, residents of the East End received 'little more than crumbs'.

- *The 'age of participation': 1970s and early 1980s*
  This period was influenced by post-1968 grass-roots and social movements (e.g. feminist, youth, gay and ethnic minority activism). These groups challenged the traditional distinctions between 'high' and 'low' culture. Governments started to give cultural policy a higher priority in urban planning, and increasingly supported 'minority' activities (e.g. experimental theatre, rock bands, independent radio stations).
- *The 'age of city marketing': from the mid-1980s to present day*
  Economic development priorities began to prevail over socio-political concerns. In response to the structural crisis of many Western European economies, governments looked to cultural policies to diversify the local economic base and to compensate for job losses in the manufacturing industries. Emphasis was placed on external image

enhancement, the attraction of inward investment, and the establishment of public–private partnerships.

(Adapted from Bianchini, 1999)

It is clear that from a cultural policy perspective, priorities have changed over the past few decades depending on the prevailing social and economic climate of the time. From the 1990s onwards, urban regeneration strategies have tended to prioritise environmental issues and the concept of sustainable development. Present-day strategies are also once again starting to recognise the needs of local communities, especially minority groups. Indeed, as stated earlier, it is evident that urban policies should be fully integrated if a successful approach to sustainable cultural planning is to be implemented.

In its 1999 Report *Towards an Urban Renaissance*, the DETR Urban Task Force emphasised the need for the right social and economic policies, combined with the right political leadership. They cite the following as being the most important factors for a sustainable urban renaissance:

- *Compact form*: recycling and economic use of land; avoiding urban sprawl;
- *Diversity*: promoting cultural innovation and community participation;
- *Connectivity*: facilitating easy movement and improving transport links;
- *Economic strength*: targeted use of public and private investment;
- *Ecological awareness*: efficient use of energy and environmental protection;
- *Good governance*: a commitment to strategic planning and participation;
- *Social inclusion*: mixed use development, and ensuring access to housing, education, health and cultural services;
- *Good design*: attractive architecture, landscaping and public space management.

(Adapted from DETR, 1999)

Urban planning is becoming more democratic, inclusive and participatory. This is reflected in cultural policy, which is aiming increasingly to maximise access to local communities and visitors alike. Bianchini (1999) shows that the concept of cultural planning has adopted a broad definition of cultural resources, consisting of:

- Arts and media activities and institutions;
- The cultures of youth, ethnic minorities and other 'communities of interest';
- The heritage, including archaeology, architecture, gastronomy, local dialects and rituals;
- Local and external perceptions of a place, as expressed in jokes, songs, literature, myths, tourist guides, media coverage and conventional wisdom;
- The natural and built environment, including public and open spaces;
- The diversity and quality of leisure, cultural, eating, drinking, and entertainment facilities and activities;
- The repertoire of local products and skills in the crafts, manufacturing and services.

The rest of this chapter will focus on the diverse ways in which culture has been used in urban regeneration strategies, as a catalyst for socio-economic development, a tourism development tool and a means of enhancing external image.

## The role of tourism, leisure and culture in urban regeneration strategies

Many cities create integrated 'cultural districts' or 'cultural quarters' as part of regeneration strategies. Wynne (1992: 19) describes a 'cultural quarter' as being 'that geographical area which contains the highest concentration of cultural and entertainment facilities in a city or town'. In many cases these will be mixed-use development areas, which aim to provide attractions and facilities for locals and tourists alike. These may include entertainment facilities, retail outlets, and eating and drinking establishments, as well as cultural venues or attractions (e.g. museums, galleries or theatres). It is important in any cultural regeneration project that cultural developments are integrated into mixed-use districts designated also for office space, residential, hotel, catering, retail and recreational use, rather than constructing isolated arts centres or cultural landmarks which fail to generate further economic and social benefits for the local communities.

Sudjic (1999) describes the growing trend in cities to redefine themselves through 'grands projets' in the French tradition. This may include the building of conference centres, universities, new airports, or, equally, cultural attractions such as museums or festivals. Bilbao, for example, has managed to transform itself from a little-known northern industrial Spanish city into the cultural centre of the Basque country and increasingly popular with the European short-break market. This is all a direct result of its 'flagship' museum The Guggenheim.

Some cities use cultural flagship projects or events as catalysts for the environmental, social, economic and cultural regeneration of an area. This may include the development of new cultural attractions, such as museums, galleries or theatres, or cultural events, such as festivals and exhibitions. It may equally include waterfront development schemes. Richards (1999a) analyses the way in which the European City of Culture initiative has been used as a tool for urban redevelopment, especially in cities that are not conventional cultural centres or cultural capitals. The example of Glasgow is often given to illustrate the success of this event. Glasgow managed to 'aestheticise' the city, enhance its external image, attract investment and encourage tourism. Bianchini (1993: 16) gives examples of other European cities such as Barcelona, Frankfurt, Liverpool and Montpellier, which have all used cultural flagships as 'powerful symbols of urban renaissance'. The contribution of cultural flagship projects and 'mega-events' to urban regeneration is also becoming increasingly well documented. Jacobs (1996: 4) sees mega-developments, heritage designations and spectacles of consumption as urban transformations which are 'hallmarks of postmodernity'. Such initiatives can act as a catalyst for numerous social, economic and cultural impacts, as described by Knox (1993: 10):

> Spectacular local projects such as downtown shopping malls, festival market places, new stadia, theme parks, and conference centers are seen as having the greatest capacity to enhance property values and generate retail turnover and employment growth.

Cities clamour to host the next Olympic Games, the Expo or the European City of Culture initiative. The legacy of 'mega-events' and their contribution to urban regeneration (such as Olympic Games or Expos) is a subject that has concerned a number of researchers (e.g. Hall, 1992a; Evans, 1996; Getz, 1997; Smith and Jenner, 1998). Hall (1992a) emphasises the importance of the economic and social impacts of such events on the host community, who are often overlooked. Sudjic (1999) suggests that cities and governments are unfortunately not very good at hosting mega-events. The legacy is often one of high levels of debt, redundant buildings and a community that has been displaced or bypassed. However, they are high-profile events which generate international publicity; therefore the host city often considers the prestige to outweigh other considerations.

The case studies outlined in Boxes 9.2 to 9.4 give examples of some of the most recent flagship mega-events and their legacies for the host cities.

## Box 9.2 Lisbon Expo 1998

Lisbon hosted the 1998 Expo, the last world exposition of the twentieth century. This is deemed a highly prestigious event, and in the past it has often served as a major catalyst for urban development and regeneration.

Portugal was the recipient of a number of EU grants in the 1990s to help stimulate economic growth. Although Portugal was one of the founder members of the common European currency, it has typically lagged behind other European countries, being the second poorest nation in the EU after Greece. The hosting of the Expo was therefore a considerable boost for Portugal in terms of economic development. It was also hoped that the development might help to salvage Portugal's ailing tourism industry. The organisers estimated that the fair would attract 8.3 million visitors, half of them from abroad, generating an additional $900 million in revenue over the year.

The 1998 Lisbon Expo marked the 500th anniversary of Vasco da Gama's discovery of the sea route from Europe to India. It also became a flagship for one of Europe's largest urban regeneration projects. Although the exposition itself lasted for only four months, its impact

**Plate 9.1** A quiet day at Lisbon's Expo site post-Expo

Source: Author

will be felt for generations. The Expo site was a rundown, polluted industrial area in East Lisbon on the banks of the River Tagus, traditionally a grimy zone of oil refineries, factories and warehouses dating back to the turn of the century. The site represented only one-fifth of the area that was developed as the riverside city, Expo Urbe. This was a mixed-use development with offices, retail and residential areas, shopping, parks, event facilities and other infrastructure improvements, including Lisbon's largest transportation hub. In addition, the Vasco da Gama bridge was built, and is currently Europe's longest fixed river crossing.

Following the closure of the exposition, the city embarked on a ten-year regeneration plan constructing the Parque das Nacoes which, by 2010, will be a million square feet of residential development, 1.5 million square feet of retail shopping, and more than three miles of open space along the Tagus River front. This is the biggest regeneration project in Lisbon since 1755 when the city was rebuilt after being demolished by an earthquake and a tidal wave.

The $2 billion Expo helped to boost economic development, acted as a catalyst for urban regeneration, and served as an important means of projecting a positive international image for Portugal. Local pride in the event appeared to be significant, rivalling football as the passionate topic of conversation. The unofficial but national slogan became 'All roads lead to Expo'. The long-term legacy of the event is thought to be significant. As stated by the Minister Antonio Costa, who was overseeing the project:

> We don't want Expo 98 to burn brightly and then go out, like a match. We want it to be a long-lasting urban project, an affirmation of Portugal and an important contribution to the economy.

Sources: Adapted from Financial Times (2001a); Lisbon Expo '98 (2001); Hatton (2001)

## The regeneration of industrial cities

Many industrial cities, particularly in North America and Northern Europe, have suffered significant economic decline as a result of de-industrialisation. Many of these cities lost their traditional industries during the 1970s and 1980s, resulting in a massive economic downturn and significant job losses. The subsequent shift towards a services-based economy has often been a slow process, especially as many industrial cities were far from being tourist attractions or cultural centres. The majority were run-down, unaesthetic, and lacking in distinctive architectural features or landmarks. The shift from a locally or regionally based manufacturing industry to a multinational service-based economy is by no means a straightforward transition.

In Britain and most other Western European countries, there was an emergence of 'arts-led' urban regeneration strategies. It was hoped that such strategies would serve to reconstruct the city's external image, making it attractive to potential investors and visitors, as well as triggering a process of physical and environmental revitalisation. Inspiration was drawn from the American experience of the 1970s in cities like Pittsburgh, Boston and Baltimore. Mayors in these cities were keen to relaunch the image of downtown areas, develop cultural districts and mixed-use areas, and to boost the local economy through arts-related activities (Bianchini, 1993). Jacobs (1961: 386) emphasised the importance of a thriving cultural life for American cities: 'We need art, in the arrangement of cities as well as in other realms of life, to help explain life to us, to show us meanings, to illuminate the relationship between the life that each of us embodies and the life outside us.' She later adds that 'lively, diverse, intense cities contain the seeds of their own regeneration' (ibid.: 462).

## Box 9.3 The Sydney Olympics 2000

It is always difficult to gauge the impacts of the Olympic Games, but Sydney was generally lauded as being something of a success story, at least in terms of the event itself. Indeed, the Olympic President Juan Antonio Samaranch declared that it had been the 'best ever Games'. Sydney spent A$3.3 billion on the Games, the supporting infrastructure and facilities. Around 700,000 visitors were expected during the Olympic period with an estimated 111,000 of those coming from overseas. The long-term external image boost was seen as being significant. Bob Carr, the New South Wales Premier, stated that it would give the city 'a great leap in tourism and convention tourism'. Indeed, for some time Australia became the number one destination for American tourists.

Jones Lang LaSalle and LaSalle Investment Management carried out an impact analysis of the event, and concluded that it had been successful in terms of improving infrastructure (e.g. transport, telecommunications, environmental initiatives), increasing international profile, and encouraging tourism and hotel development. It had also had a significant longer term impact on the real estate market, and acted as a catalyst for extensive residential, business and retail developments. The report attributes the success of the Olympics (or other world events) as catalysts for regeneration and the creation of a lasting legacy for the host city to the following factors:

- Competitiveness of the business environment, which affects the ability to attract corporate occupiers.
- Quality of tourist attractions, which determines the degree of long-term tourism benefits.
- Ability to sell Olympic experience to attract other major world events, which extends to the reuse of facilities and the leveraging of organisational experience.
- Levels of tourism infrastructure built for the Olympics, which has major long-term implications.
- Presence of an ongoing promotional campaign, which is crucial in translating the short-term interest into long-term benefits.

There are some concerns over the future of the stadium itself, which might become something of a 'white elephant', according to Anthony Hughes of the Centre for Olympic Studies at the University of New South Wales. However, the Homebush site was previously a mix of unusable swamp and a variety of semi-urban utilities including a brick works, an abattoir and a munitions dump. The area is now one of the most accessible locations in Sydney and boasts one of the world's leading sporting venues, and is one of the largest and best serviced new residential communities in Sydney. Sydney's approach was also considered to be environmentally friendly and largely sustainable, and the event was promoted as the 'Green Games'. For example, the Olympic Village now houses the world's largest solar-powered neighbourhood.

From a community perspective, the Games aimed for a reconciliation of sorts between white Australia and its indigenous population, with Aboriginal Cathy Freeman bearing the torch. The Olympic Committee was keen to focus on 'getting the politics right', stating that community needs should take precedence over the profitability of any developers.

Sources: Adapted from Hotel Online (2001); Sydney Olympics 2000 (2001a); Sydney Olympics 2000 (2001b); Joneslanglasalle (2001)

## Box 9.4 The Millennium Dome

The infamous Millennium Dome was the flagship in the redevelopment of a derelict brownfield site in southeast London, the Greenwich Peninsula. It was only in 1995 that the Millennium Commission began the work of selecting a site for their proposed millennium celebrations. The final choice was between extending the existing site at Birmingham's National Exhibition Centre, or the redevelopment of a 294-acre derelict brownfield site at Greenwich. The decision went in favour of Greenwich on the basis of the connection with Greenwich Mean Time, the belief that London would be able to attract bigger international coverage and visitors, and 'because it had the potential for a spectacular development which would at the same time regenerate a deprived area' (Smith and Jenner, 1998: 86). There was a clear political influence in the decision to use the millennium festival as a catalyst for the rejuvenation of this deprived area; as stated by Greenwich MP Nick Raynsford: 'Picking Greenwich had enormous advantages. There was a huge regeneration potential for a very depressed part of London in an area of serious industrial decline and high unemployment' (quoted in Irvine, 1999: 48).

The idea of the Dome as a way of celebrating the millennium was a legacy from the Conservative administration and, on taking office in 1997 Tony Blair's New Labour government was initially sceptical about the cost of the scheme, the use of public money and the overall value of the project. The final decision to go ahead with the construction of the Dome appeared to be influenced partly by the views of the Chairman of the British Tourist Authority David Quarmby, and Robert Ayling, the (then) Chairman of British Airways, a sponsor of the Dome. This implies that the potential attraction of tourists to the Dome was

**Plate 9.2** Controversial visitor attraction to be given new lease of life as a catalyst for regeneration

Source: Author

continued

viewed as a major consideration in its construction. However, significant problems arose due to the inadequate planning period prior to the event, the failure to apply creative vision to the actual content of the Dome, and massive overspending on the construction and operational costs.

From the outset, public and media reaction to the Dome was sceptical. As stated by Irvine (1999: 2): 'Most major building projects provoke debate and controversy, but the construction of few buildings in modern times has attracted as much scrutiny and hostility as the Millennium Dome in Greenwich.' Part of this negativity was due to the huge costs involved and the perception that the money could have been better spent elsewhere.

It is now a well-documented fact that the Millennium Dome failed to attract its (arguably ambitious) visitor target numbers (12 million), despite becoming the most popular visitor attraction in Europe during 2000. These problems notwithstanding, the Millennium Dome has acted as an important catalyst for the regeneration of the local area. This includes the development of the local infrastructure, residential developments, retail and leisure provision, employment creation for local people, education and training programmes, tourism development, and a number of ethnically diverse cultural initiatives which formed part of the Greenwich Millennium Festival. As stated by Lord Falconer (1999: 1):

> Greenwich has become a globally recognised model for regeneration, regeneration for which the Dome has been a catalyst and is helping to bring prosperity to one of the poorest parts of London and to the UK as a whole.

The long-term fate of the Dome remains to be seen, but the decision to convert it into a sports and leisure arena, coupled with the development of the Peninsula into a thriving residential 'village', is arguably a positive decision for the future regeneration of this long neglected area.

Source: Adapted from Smith and Smith (2000)

Similarly, Fisher and Owen (1991: 2) state that 'the arts are an essential part of a city's identity, neither more nor less important than other civic services'.

It is true that arts activities make a positive contribution to community life, attracting people to an area, hence improving safety on the streets, and creating a lively ambience. However, as discussed earlier in the chapter, they can also exacerbate the gentrification phenomenon by raising property prices, as the short case study outlined in Box 9.5 shows.

Parkinson and Judd (1988: 24) came to the conclusion that 'urban revitalisation in the United States increases societal inequalities'; however, this development is not unique to the USA. For example, Bianchini (1999) refers to the gentrification of the museums quarter in Frankfurt, which displaced local residents and facilities, and increased land values, rents and the local cost of living. As in New York, this process drove out many cultural producers who could no longer afford to be based within this 'cultural' district. Many other European cities have also become increasingly polarised. This is a result partly of gentrification, but also the seemingly inevitable consequence of global capitalism. Pacione (1997) describes how cities in Britain have become marked by deep social, economic and spatial division, resulting in the concept of the 'dual city', the 'polarized city' or the 'two-speed city'. Many communities have become disempowered and disenfranchised. As discussed earlier in the chapter, there is a clear need for regeneration schemes to focus on the people that they purport to benefit, namely local communities.

Walsh (1992: 135) is cynical about the potential benefits of cultural regeneration, particularly in industrial cities, stating that 'the heritagization of space in deprived regions is not designed to provide locals with cultural services, but rather to wallpaper over the

## Box 9.5 Gentrification, culture and community in New York's Lower East Side

New York's Lower East Side has always been a socially and culturally diverse district with a large immigrant community. It has also been the progenitor of many New York intellectuals, such as writers, poets, musicians, movie or theatre directors. Many artists made it their home, residing typically in loft apartments or studios (see Zukin, 1988). Its bohemian ambience made it an attractive setting for movies and novels, and it was frequently romanticised despite its pockets of social and economic deprivation. The artistic influx began in the 1970s as considerable investment was poured into art galleries, studios, dance clubs and bars. The concomitant gentrification of the district was inevitable as the area became more attractive to New York's wealthier citizens; hence rents and property prices soared. The subsequent displacement, fragmentation and marginalisation of many of the area's struggling artists and ethnic communities eventually led to the area becoming a central battleground by the late 1980s. Like many other cities where public discontent has eventually led to uprisings and riots, this district of New York suffered the same fate. It has even been suggested that the situation of deprived ethnic communities in gentrified areas of American cities like this one is akin to the plight of the native American Indians and their struggles over territory.

Source: Adapted from Smith (1991)

cracks of inner city decay in an attempt to attract revenue of one sort or another'. He cites Liverpool as an example of a city that has allowed the 'heritagization' of its public spaces to erode any sense of local heritage and identity. He describes the Albert Dock as 'a de-historicized place', 'a contrived place', and a form of 'ersatz-tourism' (1992: 143–4), where visitors do not encounter the real Liverpool or real people because they have all been displaced. The local sense of place has been eroded in order to provide a sanitised leisure experience that fails to capture the significance of Liverpool's industrial heritage and community life. Nevertheless, he concedes that although Merseyside has not attracted as much inward investment as it would have liked, there is no denying that the economic benefits of tourism have been significant.

Bianchini (1993) is more positive about the cultural regeneration process in industrial cities. He cites numerous examples of northern industrial cities in the UK and other parts of Western Europe that have benefited from cultural regeneration initiatives (e.g. Liverpool, Glasgow, Birmingham, Sheffield, Bradford, Newcastle, Swansea, Rotterdam, Bilbao, Hamburg). Many of these cities have focused on cultural flagship projects, garden festivals, 'mega-events' or waterfront development schemes. Some of these examples will be discussed in more detail in the following sections.

## Waterfront development schemes

Edwards (1996) describes the way in which abandoned docklands and waterfronts have been redeveloped in many UK cities. Again, the American model of cities such as Baltimore was often used as inspiration for UK developments. Edwards (1996) cites Jones (1993) who estimated in 1989 that there were 221 waterfront development schemes in the process of completion in the UK. Forty-one per cent of these projects had a tourism/leisure function, with this figure rising to 88 per cent in Wales. Edwards focuses in particular on Swansea, which has been developing its Maritime Quarter since the 1970s. Derelict dock basins, redundant buildings and old quay sides have been transformed into a 'marina waterfront village'. This includes a leisure centre, museums, commercial facilities, shops, cafés, bars, restaurants, art galleries, permanent and visiting historic ships and a marina. Cardiff is another interesting example of a city that is redeveloping its waterfront using retail, leisure and cultural facilities, in addition to its new flagship sports stadium. The development appears to be popular with locals and visitors alike. In other parts of Europe, industrial cities such as Rotterdam also drew on the experience of waterfront developments in cities like Baltimore and Boston. From the mid-1980s onwards it was recognised that there was a need to diversify the economic base of the city away from harbour activities, and to focus on the role of the service sector, cultural industries and tourism. The Binnenstadsplan 1985 to 1987 focused on developing the city centre for shopping and leisure, a Cultural Triangle was developed for culture and recreation, and a Waterstad for maritime tourism (Hajer, 1993). Rotterdam is now perhaps better known for its striking architecture and art museums than for its industrial maritime past.

However, Walsh (1992: 136) is disparaging of such schemes, stating that '[they] are concerned with "tarting up" space; there is in fact very little difference between one waterfront scheme and the next.' This implies that many regeneration schemes have become standardised, thus compromising the local character and identity of an area. Edwards (1996: 93) similarly admits that many waterfront developments have been criticised because of their 'poor design, lack of character and generally unimpressive environments'. This may well be true of some destinations; however, it is evident that the consequences of such schemes have also been largely positive in terms of local impact. This includes employment creation, business development, the attraction of inward investment, and the improvement of local leisure and retail provision, not to mention the attraction of visitors and the enhancement of image.

## Town centres and cultural quarters

A large number of industrial cities have made a concerted effort to develop cultural flagship projects and cultural quarters in an attempt to encourage investment, enhance local leisure, cultural and tourism provision, and to attract visitors. McGuigan (1996: 104) differentiates between American cities which place more emphasis on out-of-town retailing and leisure complexes, and the European model which is 'associated with café society. The city centre as a convivial and safe place to visit in the evenings for all sections of the population.' The city centre in European cities is usually a social meeting place, the hub of transport and communications, and cultural quarters are often mixed-use development areas with shops, cafés, bars, restaurants and entertainment venues. Some UK cities followed the American model in the 1980s, privatising public space and creating out-of-town shopping malls and entertainment complexes. This often led to underuse of town centres which adversely affected trade. The recession in the early 1990s exacerbated this problem. In his town centre research, Worpole (1992) attributed the decline of many town centres in Britain to the standardisation of retail provision, an underdeveloped evening economy, lack of adequate security provision and loss of identity. Whereas many American cities had developed twenty-four-hour facilities, and other European cities kept their retail outlets open throughout the evening, British cities typically suffered from (and still do) a dead period between shop closing hours and evening opening in bars and restaurants. The streets are perceived as lacking animation, atmosphere, often deserted and therefore deemed unsafe. The existence of arts activities such as street theatre and buskers can help to remedy this problem, but they are still a rare sight in the majority of UK town centres outside of working hours. The architect Le Corbusier declared that we should 'kill the streets', but Worpole (1992) argues that the streets are central to civic life and identity. Edensor (1998: 213) echoes this, stating that 'in most Western streets, notably high streets which were previously symbolic spaces for the production and transmission of local identity, their reconstruction or disappearance had resulted in the erasure of much social, sensual and rhythmic diversity in urban space'.

The streets are clearly places where people engage in the assertion and expression of their culture and identity. A good example of this is the development of gay quarters in UK cities. Taylor *et al.* (1996) describe how Manchester established its Gay Village in the early 1990s in pursuit of the 'Pink Pound' in an attempt to compensate for the recession. Erstwhile 'straight' venues were turned into gay ones, and the area was 'claimed' by the gay community who had previously attended low-key gay venues dispersed throughout the city. The Gay Pride event or Mardi Gras is being used increasingly as a means of asserting gay identity, but also of attracting visitors. Manchester and Brighton Gay Pride events are becoming almost as popular as London's, and cities like Birmingham, Leicester and Sheffield are following suit. However, it is interesting to note that many gay areas of cities are becoming increasingly gentrified. This is certainly true of London's Compton Street, Brighton's arcades, Manchester's 'Gay Village' and Birmingham's newly established gay quarter. There are also concerns among the homosexual community about the 'de-gaying' of certain spaces as heterosexuals are becoming more frequent visitors to such areas and venues.

## Cultural flagship projects and image enhancement in industrial cities

The use of cultural flagship projects in industrial cities became a common phenomenon during the 1980s and 1990s, especially in the north of England. Bianchini (1993) cites many examples of cities that have used museums, galleries, festivals, special events, conference centres and other cultural developments to boost their economies and image. It is clear that many industrial cities often lack the architectural and aesthetic charm of historic cities. They do not always have sufficient attractions, landmarks or cultural icons to spearhead their marketing and promotional campaigns. Such cities therefore have to become creative, innovative, and, above all, unique, in their attempt to attract inward investment and visitors alike. Where there are few natural or cultural resources, they have to create new attractions or 'flagship' developments.

For example, Glasgow opened its Burrell Collection in 1983, launched its Mayfest Arts Festival, and the successful marketing campaign 'Glasgow's Miles Better'. It later became European City of Culture in 1990. Birmingham used the construction of its International Convention Centre (ICC) in 1991 to attract prestigious orchestras, opera and ballet companies to the city, as well as developing business and conference tourism. Efforts were made to enhance the city's profile as a cultural destination by developing four 'cultural districts' around the ICC, including a Media Zone, Jewellery Quarter, theatre and entertainment district, and Chinatown. Bradford also opened the National Museum of Photography, Film and Television in 1983 in a redundant theatre building, which has acted as an important catalyst for cultural and tourism development. The Alhambra Theatre was the second flagship, and ethnic festivals also featured in Bradford's regeneration strategy. Sheffield City Council developed a cultural industries strategy and established a 'Cultural Industries Quarter' in the early 1990s. The Cultural Industries Quarter focused largely on media, television, pop music and film. This was part of a broader urban regeneration development for which the World Student Games had acted as a catalyst in 1991. Although this event was not deemed entirely successful in itself, the legacy to Sheffield was an excellent sports and leisure complex. In 1987 Liverpool City Council produced its innovative *Arts and Cultural Industries Strategy* document, which linked cultural policy with local tourism, city centre and economic development strategies. By 1988 the city had transformed the Albert Docks into a spectacular arts, leisure and retail complex, including the Tate Gallery, a Maritime Museum, and Granada TV's regional news centre and offices. The legacy of the Beatles and the success of the Liverpool-based British soap opera *Brookside* helped to further enhance the city's image. Newcastle is another city that developed a cultural regeneration policy in the late 1980s to assert the function of Newcastle as a regional capital. The City Council helped renovate and relaunch a number of arts venues, including the Theatre Royal, the Tyne Theatre and Opera House, and the Tyneside Cinema, as well as the establishment of Newcastle Arts Centre.

However, one of the most successful cultural flagship developments in Europe in recent years has been the construction of the Guggenheim Museum in Bilbao, as the case study outlined in Box 9.6 will demonstrate.

## Box 9.6 Bilbao and the Guggenheim

Bilbao is an interesting example of a European industrial city that has used one cultural flagship project, the architecturally spectacular Guggenheim Art Museum, as a major catalyst for regeneration and image enhancement. Richards (1999a) comments on the 'raised eyebrows' that greeted the decision to locate Gehry's cultural icon in such an unlikely location. Sudjic (1999: 181) suggests that the location was carefully and deliberately chosen: 'Obviously the Guggenheim didn't open in Bilbao for the sake of it; it was an attention grabber. It was part of a wholesale reconstruction of the city.'

In the early 1990s, Bilbao was struggling against the demise of its traditional industries. Shipyards and steel mills had closed, unemployment was high and Basque terrorism was rife. During the 1990s, prior to the construction of the Guggenheim, the city's redevelopment plans included a new subway system, the reconstruction of the city's airport, a waterfront development and the Uribitarte Footbridge. La Vieja district was regenerated as part of an EU-funded wider General Urban Development plan, which focused on infrastructure, accessibility, and physical, social and economic rehabilitation.

There were concerns about the proposed construction of a museum to spearhead the city's revival, especially given budgetary constraints. However, the impact of the Guggenheim has surpassed the city authorities' expectations. It has recouped its $76 million investment, and it is estimated that 3.5 million tourists have visited the attraction, injecting $400 million into Bilbao's economy. According to the museum's director, Juan Ignacio Bidart: 'We have recovered our self-esteem. Suddenly, Bilbainos feel that it is possible to reverse the city's trajectory of industrial decline.'

This development has clearly put Bilbao on the map, and tourism development has been encouraged to one of Spain's lesser-known regions. In the context of a broader regeneration strategy where culture is just one of many strands, it serves as an attractive flagship and central focus for the cities' visitors. It is arguably the jewel in Bilbao's regeneration crown, and it is likely to set a precedent. As stated by Josu Bergara, President of the Regional Council of Vizcaya: 'Suddenly, everybody wants a Guggenheim.'

Sources: Adapted from European Commission (2001); Guggenheim Bilbao Museum (2001a); Guggenheim Bilbao Museum (2001b)

## Cultural regeneration: the production of new spaces and the (re)creation of place

Hughes (1998) describes how tourism is a spatially differentiating activity which can lead to the homogenisation of culture, but which can also help to 're-vision' or 're-imagine' space:

> Tourism . . . differentiates space in a ceaseless attempt to attract and keep its market share. In the face of growing global cultural homogenisation, local tourist agencies strive to assert their spatial distinctiveness and cultural particularities in a bid to market each place as an attractive tourist destination.
>
> (Hughes, 1998: 30)

Something similar could be said of the regeneration process. Various destinations are engaging actively in the reconfiguration of their identity in an attempt to reposition themselves or to put themselves on the tourist map. However, Walsh's (1992) fears about the bland standardisation of waterfront developments or the 'heritagization' of public space are not unfounded. It is evident that numerous town centres, particularly in Britain, are starting to rely on inward investment from global businesses, which render them at best homogeneous and at worst soulless. Although it is clear that former industrial cities often have little option but to court such investment, it can quite feasibly be channelled into the development of innovative new projects, initiatives and attractions, rather than bland retail developments.

If we return for a minute to our discussion in Chapter 1 about the motivation of the post-tourist, it is clear that increasing numbers of tourists are drawn to the excitement of 'hyper-real' experiences, often within enclavic bubbles, such as shopping malls, theme parks or leisure complexes. The production of such spaces appears to be a prominent characteristic of postmodern urban planning, and thus an inherent part of regeneration. Edensor (2001) describes the way in which tourist space has become increasingly themed, especially in areas such as Leicester Square/Piccadilly Circus in London, where the everyday and the extraordinary are mixed in attractions like Rock Circus, Sega World, Planet Hollywood and the Fashion Café.

Regeneration also helps to 'aestheticise' space through enhancement schemes. These may take the form of architectural 'face-lifts', the design of public art, or the transformation and beautification of former industrial sites. However, Hughes (1998) suggests that care must be taken not to 'overwrite' the original significance of heritage spaces in the process. Middleton (1987) comments on the problems of towns becoming too sanitised in insidious attempts to clean up and prettify, and Bianchini (2000: 5) suggests that cultural developments must be integrated and not be a mere adjunct, described as 'putting lipstick on the gorilla'.

The common characteristic both of the tourism development and regeneration process is that they seek to transform old spaces while re-creating new ones. Tourism life-cycle models often imply that old destinations rarely die; they simply rejuvenate! While it is seldom possible to rectify environmental damage and the destruction of natural resources, culture is arguably more resilient and can withstand changes of fortune. Cultures rarely disappear, they simply evolve. Notwithstanding the sweep of globalisation, many destinations are exploring innovative and creative ways of expressing or representing regional or local cultures, all of which serve as unique reminders of their individualism and identity.

## Conclusion

In many ways, the latter part of this chapter has brought us back to a recurrent theme within this book, which has looked at the way in which spaces are constantly appropriated by dominant groups. This might include the claiming of indigenous homelands in the case of imperial colonisers; or the dominance of one particular nation (e.g. America) in media and cyberspace; the way in which tourists set up enclavic spaces within a destination or 'hijack' local resources and facilities; or the way in which planners gentrify urban space. Once again, the concept of power is unavoidable in such discussions of spatial superiority.

Tourism is clearly a dynamic force and one which (like all forms of development) will engender economic, social and cultural change. While it inevitably degrades some spaces and places, it will also be instrumental in creating new ones. Although a balance must be struck between the preservation and conservation of heritage for present and future generations, those features and attractions which now strike us as new, alien or 'post-modern' will one day form part of the heritage that we will wish to preserve. Human beings are surprisingly resistant to change, but all forms of development are necessarily about transformation. We perhaps need to become even more flexible in our attitudes towards cultural change. Only then can we benefit from the rich diversity of exciting and innovative new developments while keeping half an eye on those aspects of our heritage which have made us what we are.

# Conclusion

> If our lives are dominated by a search for happiness, then perhaps few activities reveal as much about the dynamics of this quest – in all its ardours and paradoxes – than our travels. They express, however inarticulately, an understanding of what life might be about, outside the constraints of work and the struggle for survival.
>
> (de Botton, 2002: 9)

The aim of this book has been to demonstrate the complexity and diversity of the phenomenon of cultural tourism. The broad definition of culture adopted from the outset has necessitated an in-depth discussion of a variety of issues, many of which are controversial, sensitive or highly politicised. However, the philosophy underpinning this book has been that culture is a contentious concept, and one which can be simultaneously inclusive or exclusive depending on the way in which it is managed.

The globalisation of culture is discussed in some detail, particularly as tourism is the quintessential global industry. It contains all the contradictory elements of the globalisation process, being something of a double-edged sword. While bestowing its much sought-after economic benefits on destinations and their host populations, it can despoil their habitat and rob them of their traditions. Like all global industries, tourism is dominated largely by major corporations and Western tourists from developed countries. Hence it is not surprising that it has often been described as a new form of imperialism. Like colonialism, it subjects others to its power, leaving them with little choice but to succumb to its influence.

Clearly, European imperialism left its mark on the world in a variety of contexts, not least in the field of culture. Indigenous peoples the world over were generally displaced from their lands and their homes, forced to adhere to colonial rule and all its concomitant customs and traditions. Not only were native traditions usually considered to be inferior, but they were even deemed unworthy of continuation; hence they were often suppressed, discarded or forbidden. The legacy for such peoples has inevitably been one of great loss and disinheritance. It has only been in recent years that a small number of sympathetic governments, charities and action groups have been dedicated specifically to their cause. There has subsequently been, in some cases, a revival of customs and traditions, and a renewal of cultural pride. The role that cultural tourism can play in this process is a subject of ongoing debate, but this book has attempted to demonstrate the positive benefits tourism can have on indigenous populations if it is managed properly. Many of the issues relate to the furthering of local ownership and empowerment, processes that are possible only if adequate political, financial and moral support is provided.

The impacts of cultural tourism can clearly be major, especially as it could now be considered a growth industry rather than a minority pursuit. As discussed, the diversity of cultural tourism activities can give rise to a number of diverse impacts. It is evident that the impacts of tourism will always be most significant in those areas of the world that are particularly remote, fragile or unaccustomed to tourism. One of the criticisms of tourism as a new form of imperialism is that tourism flows predominantly from Western developed countries to less developed countries. The local populations are usually immobile both physically and financially, at least in touristic terms; therefore their role will never be more than that of serving tourists. They will never experience what it is like to be on the other side, as it were. Once again they are subordinated, relegated to the status of little more than servants to wealthy Westerners. It is not surprising, therefore, that resentments are common, and that tourists often find themselves victims of hostility, harassment or crime. The sweep of globalisation has merely exacerbated the situation, opening up even more destinations to those who can afford the privilege.

There are no easy solutions. Governments are frequently lured by the bait of tourism development simply because it appears to offer a 'quick-fix' solution to the country's economic difficulties, and potentially affords them the opportunity to enter the global arena. However, consideration is not always given to the long-term problems of such developments, nor is the revenue generated from tourism necessarily reinvested in the destinations and host populations. Such issues are well documented elsewhere; therefore it has not been the aim of this book to dwell on such problems in any great depth. Nevertheless, it is recognised that cultural tourism is as politicised as any other form of tourism development, and in many ways more so, because it focuses on the whole way of life of a country's people.

One of the other colonial legacies that is dealt with in some detail in this book is that of the interpretation and representation of culture through heritage and museum collections, both of which constitute an important element of the cultural tourism product. Whole books have been devoted to this subject; therefore a synthesis of the main theories is presented here. This is perhaps one of the most complex and sensitive areas of cultural tourism management. In the past there was very little recognition of minority, ethnic or indigenous cultures. As stated earlier, the history and culture of indigenous peoples were often suppressed or destroyed by colonial powers who considered them to be inferior, primitive, or even barbaric. Ironically, however, when worthwhile objects or artefacts were discovered, they were often removed and placed in European museums. The subject of ownership of cultural property is consequently still a hotly debated issue within museum studies.

Although the situation is slowly changing, it has only been since postmodern debates about the rejection of so-called 'grand narratives' came along that there has been a growing interest in the history and culture of previously marginalised or oppressed peoples. This has altered not only approaches to the study of history (which has always been predominantly a white, Western, patriarchical account of the past), but it has also affected the nature of heritage interpretation, arguably for the better. Although many critics have argued that the heritage and museums 'industries' have been over-commercialised and trivialised as a consequence, some might contend that this is merely an elitist or Eurocentric stance. The representation of the history and heritage of the working classes, women, ethnic minorities and indigenous peoples is surely a welcome development in an age when there is such a definite need to move towards more inclusive, democratic and participatory approaches to cultural development. It is clear that there are still too few ethnic or indigenous curators in many museums and galleries (hence interpretation is often carried out by those with only second-hand knowledge), but the formal recognition of this problem indicates, with any luck, a willingness to change.

Significant changes are also taking place within the heritage field as attempts are being made to diversify the concept of worthwhile heritage. Emphasis is being placed on historical value or 'historicity' rather than the mere aesthetics of sites. For example, there has been a shift away from the dominance of royal palaces, castles and country houses on the World Heritage List to the inclusion of industrial heritage sites or more intangible forms of heritage belonging to indigenous peoples or the working classes. Again, this is a welcome development, although there are some concerns that the integrity of the World Heritage List will in some way be compromised by this development. It could of course be argued that World Heritage should not only be unique, it should also be awe-inspiring. How far therefore can a former mining landscape compete with the splendour of the Taj Mahal? Again, this is a subject of ongoing debate, but the fundamental principle of adopting a more inclusive and less Eurocentric approach is surely a positive one.

The concept of access is clearly an important one for UNESCO, and part of the organisation's remit is to make World Heritage sites available to the widest possible public. However, this can bring significant problems in its wake, particularly as the inscribing of sites on the List generally raises their profile and puts them on the tourist map. Although tourism can help to provide useful funds for conservation, questions must be raised about the extent to which fragile sites can withstand the ravages of cultural tourism for all eternity. The concept of sustainability was developed partly in order to address the issue of perpetuity; that is, the consideration of heritage as being not only about the past, but also about the present and future. Clearly, there is little point in preserving the heritage for future generations if current generations cannot benefit from it in the meantime. However, the growth of international tourism is creating enormous problems, especially for some of the world's 'must-see' heritage sites, many of which are inscribed on the World Heritage List. Again, there are no easy solutions, but care must be taken to manage the sensitive balance between conservation, access and visitor management. There is also the question of the local community and their relationship to heritage sites, especially if they happen to live within it (i.e. in the case of historic towns or national parks). Issues of multiple interpretation also clearly need to be addressed.

Tourism has also been criticised for reducing the heritage and museum sectors to a mere entertainment forum. The (often severe) lack of funding for conservation and collections management has forced heritage sites and museums to diversify into other activities (e.g. tourism or retail development), often at the perceived cost of their core function. This is a serious issue, and questions should be raised about the extent to which it is acceptable for museums and heritage sites to resemble theme parks or entertainment zones. After all, there is still a place for education, particularly within the field of cultural tourism. Again, striking a balance between the education and entertainment functions of such attractions is the key challenge. It is particularly difficult in the case of 'dissonant' heritage sites to provide a form of interpretation that is sensitive to all parties. It may not be appropriate to develop tourism in these cases, and such sites should remain educational memorials.

Of course, it is not only in the heritage and museum sectors that issues relating to inclusion, access and democracy have become a major priority. They have perhaps become even more of a key issue within the arts. There has been growing concern that the arts have been traditionally elitist in their focus on so-called high culture and their dismissal of popular or mass culture. Although many mass forms of culture (e.g. pop music, fashion) can generate their own income much more easily than high arts which invariably need subsidies, it is more a question of prevailing attitudes that need to be changed. The post-colonial legacy has led to increasing multiculturalism and ethnic diversity in many societies; hence the culture and arts of such groups need to be given some recognition and support. Although many arts activities such as the Notting Hill Carnival have grown organically from a strong community base, their future continuation is highly dependent on funding and political support.

The relationship between tourism and the arts has often been less than harmonious, only because there is some fear within arts circles that tourism will somehow compromise the artistic integrity or authenticity of performances or events. Like the heritage and museum sectors, the arts have often been forced to adapt their core product to attract more diverse audiences, partly because of access issues but also because of lack of funding. In many cases, programming is constrained by these factors, which is a great pity.

There has also been much debate about the use of the arts for other functions; for example, as a tool for urban regeneration. Many cities have developed cultural quarters or 'flagship' projects based on the arts and related activities. It has even reached the stage where the content of a museum or gallery is somehow becoming less relevant than the building's potential as a catalyst for regeneration and the attraction of investment. Both the Guggenheim in Bilbao and Tate Modern in London have been criticised for being examples of this phenomenon. Certainly, the arts should not always be thought of as a tool for economic or business development. This is merely a legacy from governments which are basically unwilling to subsidise the arts, forcing them into a position where they have to rely on business sponsorship or tourism in order to survive. It is inevitable that artistic content will somehow be compromised along the way.

This book has considered in some detail the artistic events and festivals of ethnic and minority groups, many of which are community-based. Although large tourist audiences are not always welcome at such events, it can help to raise their international status and provide the communities with an enhanced profile, and ultimately, the possibility of increased political power. From the tourists' perspective, such events are an increasingly attractive part of the cultural tourism product. There are few barriers to access, many such events are free, they are participatory, and they afford the participant a glimpse of authentic, indigenous or ethnic culture. Of course, if the events become less spontaneous and 'staged' for the benefit of tourists, then they will inevitably lose their authenticity and ultimately their appeal. However, it appears to be the case with festivals and carnivals that such events are still largely spontaneous, mainly because they are usually a celebration of a particular phenomenon or moment in time.

Some of the recurrent themes in this book have been issues relating to integration and identity. Almost every chapter refers to these concepts to a greater or lesser extent. Once again, these are topical themes in a number of fields, but the role that culture plays in the development of identity is being increasingly recognised. It is clear that the identity of indigenous, minority and ethnic groups has frequently been suppressed during various periods throughout history. Even today, many groups are still unable to express themselves freely, mainly because of oppressive political regimes, but even in so-called democracies, because of prejudice, intolerance and racism. Chapter 4 discussed these issues in the European context, focusing mainly on the cultural initiatives that have been developed in recent years in order to further the process of integration and the strengthening of identities.

The globalisation process has clearly had a major influence on both integration and identity. As stated earlier, many societies have become much more culturally and ethnically diverse since decolonisation. As a result, although cities have become more cosmopolitan, they have also become more polarised, and in some cases, more ghettoised. This is perhaps largely to do with the inadequacy of politics and economics to resolve issues of integration. For example, it was evident that the traditional focus of the EU on political and economic unification did little to further the process of social or cultural integration within Europe. However, efforts by cultural and arts organisations to enhance cross-cultural exchange and to develop networks and joint initiatives has arguably been more effective in bringing people together from different communities, cultures and races.

Identity is a rather fluid concept, especially in the postmodern, global era. Again, numerous books and articles have been devoted to this phenomenon. Identity is surely something to do with the way we feel either individually or collectively, rather than something that is imposed upon us. Of course, under an oppressive political regime, any personal sense of identity is likely to be suppressed. However, even in democratic societies, many writers have referred recently to the apparent identity crisis that seems to be plaguing many people within Western, global societies. Increasing mobility, the fragmentation of communities and the family unit, and the decline of religion have clearly contributed to this. But so too has the globalisation process itself. The development of mass media, high technology, instantaneous communications and faster transport has made the world a smaller place, but it has also led to a creeping standardisation or homogeneity. Many tourists are now travelling to seek difference outside the home as their everyday lives have become an integral part of the standardised global process. Although the world of cyberspace has afforded people the capacity to escape virtually at any time, paradoxically this has led to a marked increase rather than a decrease in the demand for tourism.

Tourism and cultural tourism are clearly still growth industries, despite the assumption that the home entertainment revolution would increase people's desire to stay at home. Cynics might argue that we are less than happy with the world that we have created for ourselves, and are only too keen to escape at the earliest opportunity and 'get away from it all'. It is no coincidence that stress levels are rising as working hours increase, and we are becoming more and more suffocated by technology and communications, the monsters of our own making. The burgeoning alternative medicine and spiritual healing industries are perhaps indicative of the crisis that is dominating global Western societies. It is ironic that many developing societies are hankering after the same economic 'progress' and lifestyle that many Westerners are now so keen to escape.

Conversely, it is no surprise that many indigenous communities are so keen to preserve their culture and traditions which involve being at one with nature and at peace with themselves (if only they were left alone to get on with it). It is also no surprise that increasing numbers of Western tourists are embarking on trips in which they 'get back to nature', or engage in some kind of spiritual quest. Cultural tourism affords us the ability to be who we want to be, if only for a while. As stated at other points in this book, tourism is so much about the production of dreams, the indulgence in fantasies and escapism into ideal worlds.

Unfortunately, wealthy Western societies have argubly not created the ideal world for themselves in terms of their spiritual and mental well-being. Such societies are also still as polarised as they ever were, both economically and culturally, and arguably they have a long way to go before they become truly democratic and inclusive. In other parts of the world, many communities are barely subsistent. Abandoned or oppressed by their governments, tourism often appears like the elusive coffer of gold at the end of a rainbow. Of course, if managed well, cultural tourism can become something of a panacea – not just for the wealthy Westerner, keen to experience cultural difference, or in need of physical, mental or spiritual rejuvenation; but also for those societies that really need an alternative economic development option in order to survive. Thus culture may be ordinary, but cultural tourism arguably has the potential to become truly extraordinary.

# Bibliography


Adair, G. (1992) *The Postmodernist Always Rings Twice: Reflections on Culture in the 90s*, London: Fourth Estate.

Adams, R. (1986) *A Book of British Music Festivals*, London: Robert Royce.

Addison, L. (1996) 'An approach to community-based tourism planning in the Baffin region, Canada's Far North', in Harrison, L. and Husbands, W. (eds) *Practising Responsible Tourism: International Case Studies in Tourism Planning, Policy and Development*, Toronto: John Wiley and Sons, pp. 296–312.

Adorno, T.W. (2001) *The Culture Industry: Selected Essays on Mass Culture*, London: Routledge.

Adorno, T.W. and Horkheimer, M. (1979) *Dialectic of Enlightenment*, London: Verso (first published in 1947).

Ali, Y. (1991) 'Echoes of empire: towards a politics of representation', in Corner, J. and Harvey, S. (eds) *Enterprise and Heritage: Crosscurrents of National Culture*, London: Routledge, pp. 194–211.

Alleyne-Dettmers, P.T. (1996) *Carnival: The Historical Legacy*, London: Arts Council of England.

AlSayyad, N. (2001) 'Global norms and urban forms in the age of tourism: manufacturing heritage, consuming tradition', in AlSayyad, N. (ed.) *Consuming Tradition, Manufacturing Heritage: Global Norms and Urban Forms in the Age of Tourism*, London: Routledge, pp. 1–33.

Anderson, C. (1995) *Our Man In . . .*, London: BBC Books.

Anderson, M.J. (1991) 'Problems with tourism development in Canada's Eastern Arctic', *Tourism Management*, 12 (3), pp. 209–220.

Appadurai, A. (1990) 'Disjuncture and difference in the global cultural economy', in Featherstone, M. (ed.) *Global Culture: Nationalism, Globalisation and Modernity*, London: Sage, pp. 295–310.

Arnold, M. (1875) *Culture and Anarchy*, London: Smith, Elder.

Art Cities in Europe (2002) (online) http://www.artcities.de/euns/allaboutus.html (accessed 3 March).

Arts Council of Great Britain (1986) *Masquerading: The Art of the Notting Hill Carnival*, London: ACGB.

Ashworth, G. (1995) 'Heritage, tourism and Europe: a European future for a European past?', in Herbert, D.T. (ed.) *Heritage, Tourism and Society*, London: Mansell Publishing, pp. 68–84.

Atkinson, R. (2000) 'Measuring gentrification and displacement in Greater London', *Urban Studies*, 37 (1), pp. 149–165.

Baig, A. (2001) (online) *Agra Fort – Conservation of Heritage in the Urban Context*, http://www.tobunken.go.jp/~kokusen/japanese/7SEMINAR/baig.html (accessed 26 September).

Bakhtin, M. (1965) *Rabelais and his World*, Cambridge: MIT Press.

Barber, B. (1995) *Jihad vs the World*, New York: Times Books.

Barnett, S. (1997) 'Maori tourism', *Tourism Management*, 18(7), pp. 471–473.

Baudrillard, J. (1988) *Selected Writings*, ed. M. Poster, Cambridge: Polity Press.

Bauman, Z. (1998) *Globalization: The Human Consequences*, Oxford: Blackwell.

Bayles, M. (1999) 'Tubular nonsense: how not to criticise television', in Melzer, A.M., Weinberger, J. and Zinman, M.R. (eds) *Democracy and the Arts*, Ithaca and London: Cornell University Press, pp. 159–171.

Beddoe, C. (1998) 'Child abuse: is it a tourism industry issue?', in *Embodied Commodities: Sex and Tourism*, Tourism Concern, *In Focus*, 30 (winter), pp. 16 17.

Beynon, J. and Dunkerley, D. (eds) (2000) *Globalization: The Reader*, London: The Athlone Press.

Berry, J. (2001) (online) *Outrage at Fast Food, Cyber Cafe Plan for Taj Mahal*, http://www.theage.com.au/news/world/2001/07/10/FFX0LJO4XOC.html (accessed 26 September).

Bianchini, F. (1990) 'Urban Renaissance? The arts and the urban regeneration process', in MacGregor, S. and Pimlott, B. (eds) *Tackling the Inner Cities: The 1980s Reviewed, Prospects for the 1990s*, Oxford: Clarendon Press, pp. 215–250.

Bianchini, F. (1993) 'Culture, conflict and cities: issues and prospects for the 1990s', in Bianchini, F. and Parkinson, M. (eds) *Cultural Policy and Urban Regeneration – The West European Experience*, Manchester: Manchester University Press, pp. 199–213.

Bianchini, F. (1999) 'Cultural planning for urban sustainability', in Nystrom, L. (ed.) *City and Culture: Cultural Processes and Urban Sustainability*, Kalmar: The Swedish Urban Environment Council, pp. 34–51.

Bianchini, F. (2000) 'From "cultural policy" to "cultural planning"', *Artsbusiness*, 27 March, pp. 5–6.

Bianchini, F. and Parkinson, M. (eds) (1993) *Cultural Policy and Urban Regeneration: The West European Experience*, Manchester: Manchester University Press.

Blaenavon Industrial Landscape (2001) (online) http://www.world-heritage-blaenavon.org.uk/ (accessed 18 October).

Blakey, M. (1994) 'American nationality and ethnicity in the depicted past', in Gathercole, P. and Lowenthal, D. (eds) *The Politics of the Past*, London: Routledge, pp. 38–48.

Boissevain, J. (1997) 'Problems with cultural tourism in Malta', in Fsadni, C. and Selwyn, T. (eds) *Sustainable Tourism in Mediterranean Islands and Small Cities*, London: MED-CAMPUS, pp. 19–29.

Boniface, P. (1995) *Managing Quality Cultural Tourism*, London: Routledge.

Boniface, P. and Fowler, P.J. (1993) *Heritage and Tourism in 'the Global Village'*, London: Routledge.

Boorstin, D. (1964) *The Image: A Guide to Pseudo-Events in America*, New York: Harper & Row.

Borley, L. (1995) 'Heritage and environmental management: the international perspective', in *Tourism and Culture: Global Civilisation in Change?*, Yogyakarta: International Tourism Conference Proceedings, pp. 180–188.

Botton de, A. (2002) *The Art of Travel*, London: Hamish Hamilton.

Bourdieu, P. (1984) *Distinction: A Critique of the Judgement of Taste*, London: Routledge.

Bradford, G., Gary, M. and Wallach, G. (2000) *The Politics of Culture: Policy Perspectives for Individuals, Institutions and Communities*, New York: New Press.

Bradford Mela (2001) (online) http://www.bradford.gov.uk/ and http://www.oriental.legend.org.uk/mela.html (accessed 10 December).

British Tourist Authority (2002) *Cultural Tourism: How You Can Benefit*, London: BTA.

Brustein, R. (1999) 'Democracy and culture', in Melzer, A.M., Weinberger, J. and Zinman, M.R. (eds) *Democracy and the Arts*, Ithaca and London: Cornell University Press, pp. 11–25.

Bull, A. (1997) 'Italian regionalism and the Northern League', in *Papers from the ECTARC Convocation*, 11 July, Llangollen, Wales.

Burns, P.M. (1999) *An Introduction to Tourism and Anthropology*, London: Routledge.

Burtenshaw, D., Bateman, M. and Ashworth, G. (1991) *The European City: A Western Perspective*, London: David Fulton.

Busby, G. and Klug, J. (2001) 'Movie-induced tourism: the challenge of measurement and other issues', *Journal of Vacation Marketing*, 7(4), pp. 316–332.

Butcher, J. (ed.) (2001a) *Innovations in Cultural Tourism*, Proceedings of the 5th ATLAS International Conference, Rethymnon, Crete, 1998, Tilburg: ATLAS.

Butcher, J. (2001b) 'Cultural baggage and cultural tourism', in *Innovations in Cultural Tourism*, Proceedings of the 5th ATLAS International Conference, Rethymnon, Crete, 1998, Tilburg: ATLAS, pp. 11–18.

Butler, R. (1980) 'The concept of a tourism area cycle of evolution', *Canadian Geographer*, 24, pp. 5–12.

Butler, R. and Hinch, T. (eds) (1996) *Tourism and Indigenous Peoples*, London: International Thomson Business Press.

Bywater, M. (1993) 'The market for cultural tourism in Europe', *Travel and Tourism Analyst*, 6, pp. 30–46.

Cantor, P.A. (1999) 'Waiting for Godot and the end of history: postmodernism as a democratic aesthetic', in Melzer, A.M., Weinberger, J. and Zinman, M.R. (eds) *Democracy and the Arts*, Ithaca and London: Cornell University Press, pp. 172–192.

Carlson, M. (1996) *Performance: A Critical Introduction*, London: Routledge.

Carnegie, E. (1996) 'Trying to be an honest woman: making women's histories', in Kavanagh, G. (ed.) *Making Histories in Museums*, London: Leicester University Press, pp. 54–65.

Castells, M. (1994) 'European cities, the informational society, and the global economy', *New Left Review*, 204, pp. 18–32.

Castells, M. (1997) *The Power of Identity*, Oxford: Blackwell.

Castro-Gomez, S. (2001) 'Traditional vs. critical cultural theory', *Cultural Critique*, 49 (autumn), pp. 139–154.

CLTV (2001) (online) *The 2000 Olympic Superstars were Sydney and Australia*, http://cnews.tribune.com/news/story/0,1162,cltv-sports-78742,00.html (accessed 21 September).

Coelho, P. (1992) *The Pilgrimage*, London: HarperCollins.

Cohen, E. (1972) 'Towards a sociology of international tourism', *Social Research*, 39 (1), pp. 64–82.

Cohen, E. (1988) 'Authenticity and commoditization in tourism', *Annals of Tourism Research*, 15, pp. 371–386.

Colenutt, B. (1991) 'The London Docklands Development Corporation: has the community benefited?', in Keith, M. and Rogers, A. (eds) *Rhetoric and Reality in the Inner City*, London: Mansell Publishing, pp. 31–41.

Community Development Foundation (1995) *Guidelines to the Community Involvement of the SRB Challenge Fund*, London Borough of Greenwich, April.

Connor, S. (1989) *Postmodern Culture: An Introduction to Theories of the Contemporary*, Oxford: Blackwell.

Cooper, C., Fletcher, J., Gilbert, D., Shepherd, R. and Wanhill, S. (1998) *Tourism: Principles and Practice*, Harlow: Longman.

Council of Europe (1994) *Texts Concerning Culture at European Community Level*, Strasbourg: General Secretariat.

Council of Europe (1995) *Newsletter on the Council of Europe Cultural Routes*, No.3, Strasbourg: General Secretariat.

Council of Europe (2002) (online) http://culture.coe/tr/Inforcentre/pub/eng/erout4.3.htm (accessed 28 February).

Coupland, A. (1992) 'Docklands: dream or disaster?', in Thornley, A. (ed.) *The Crisis of London*, London: Routledge, pp. 149–162.

Craik, J. (1997) 'The culture of tourism', in Rojek, C. and Urry, J. (eds) *Touring Cultures*, London: Routledge.

Creamer, H. (1990) 'Cultural resource management in Australia', in Gathercole, P. and Lowenthal, D. (eds) *The Politics of the Past*, London: Unwin Hyman, pp. 130–139.

Crick, M. (1988) 'Sun, sex, sights, savings and servility', *Criticism, Heresy and Interpretation*, 1, pp. 37–76.

Crilley, D. (1993) 'Megastructures and urban change: aesthetics, ideology and design', in Knox, P.L. (ed.) *The Restless Urban Landscape*, Englewood Cliffs, NJ: Prentice Hall: pp. 127–164.

Crowhurst Lennard, S.H. and Lennard, H.L. (1987) *Livable Cities*, Southampton: Gondolier Press.

Cummins, A. (1996) 'Making histories of African Caribbeans', in Kavanagh, G. (ed.) *Making Histories in Museums*, London: Leicester University Press, pp. 92–104.

Cunliffe, S. (1996) 'Protection through site management', *Proceedings of the International Conference on Tourism and Heritage Management*, Yogyakarta, pp. 261–271.

Dann, G. (1996a) 'The people of tourist brochures', in Selwyn, T. (ed.) *The Tourist Image: Myths and Myth Making in Tourism*, London: John Wiley & Sons, pp. 61–81.

Dann, G. (1996b) 'Images of destination people in travelogues', in Butler, R. and Hinch, T. (eds) *Tourism and Indigenous Peoples*, London: International Thomson Business Press, pp. 349–375.

Dann, G. (ed.) (2002) *The Tourist as a Metaphor of the Social World*, Wallingford: CABI.

Davies, N. (1997) *Europe: A History*, London: Pimlico.

De Kadt, E. (1994) 'Making the alternative sustainable: lessons from development for tourism', in Smith, V.L. and Eadington, W.R. (eds) *Tourism Alternatives: Potentials and Problems in the Development of Tourism*, Chichester: John Wiley & Sons, pp. 47–75.

DETR (1997) *Involving Communities in Urban and Rural Regeneration: a Guide for Practitioners*, London: Crown Copyright.

DETR Urban Task Force (1999) *Towards an Urban Renaissance*, London: E & FN Spon.

De Vidas, A.A. (1995) 'Textiles, memory and the souvenir industry in the Andes', in Lanfant, M., Allcock, J.B. and Bruner, E.M. (eds) *International Tourism: Identity and Change*, London: Sage, pp. 67–83.

Delamere, T.A., Wankel, L.M., and Hinch, T.D. (2001) 'Development of a scale to measure resident attitudes toward the social impacts of community festivals, Part 1: Item generation and purification of the measure', *Event Management*, 7, pp. 11–24.

Dimaggio, P. and Useem, M. (1978) 'Social class and arts consumption: the origin and consequences of class differences in exposure to the arts in America', *Theory and Society*, 5, pp. 141–161.

Dirlik, A. (1999) 'Is there history after Eurocentrism? Globalism, postcolonialism, and the disavowal of history', *Cultural Critique*, 42 (spring), pp. 1–34.

Douglas, N. and Douglas, N. (1996) 'Tourism in the Pacific: historical factors', in Hall, C.M. and Page, S.J. (eds) *Tourism in the Pacific: Issues and Cases*, London: International Thomson Business Press, pp. 19–35.

Doxey, G.V. (1975) *A Causation Theory of Visitor–Resident Irritants: Methodology and Research Inference*, San Diego: The Travel Research Association Conference, no. 6, pp. 195–198.

Du Gay, P. (ed.) (1997) *Production of Culture/Cultures of Production*. London: Sage.

Durrans, B. (1988) 'The future of the other: changing cultures on display in ethnographic museums', in Lumley, R. (ed.) *The Museum Time Machine*, London: Routledge, pp. 144–169.

Eagleton, T. (2000) *The Idea of Culture*, Oxford: Blackwell.

Eco, U. (1986) *Travels in Hyper-Reality*, London: Picador.

Ecotourism Society (1991) (online) http://www.ecotourism.org/ (accessed 25 October 2001).

ECTARC (1989) *Contribution to the Drafting of a Charter for Cultural Tourism (Tourism and the Environment)*, Llangollen: ECTARC.

Edensor, T. (1998) 'The culture of the Indian street', in Fyfe, N.K. (ed.) *Images of the Street: Planning, Identity and Control in Public Space*, London: Routledge, pp. 205–221.

Edensor, T. (2001) 'Performing tourism, staging tourism: (re)producing tourist space and practice', *Tourist Studies*, 1 (1), pp. 59–81.

Edwards, J.A. (1996) 'Waterfronts, tourism and economic sustainability: the United Kingdom experience', in Priestley, G.K., Edwards, J.A. and Coccossis, H. (eds) *Sustainable Tourism? European Experiences*, Wallingford: CABI, pp. 86–98.

Edwards, J.A. and Llurdes, J.C. (1996) 'Mines and quarries: industrial heritage tourism', *Annals of Tourism Research*, 23, (2), pp. 341–363.

English Historic Towns Forum (EHTF) (1999) *Making the Connections: A Practical Guide to Tourism Management in Historic Towns*, Bristol: EHTF.

Errol, J. (1986) 'Mama look a Mas', in *Masquerading: The Art of the Notting Hill Carnival*, London: Arts Council of Great Britain, pp. 7–19.

European Commission (2001) (online) *Bilbao la viejz* http://europa.eu.int/comm/urban/casestudies/c010_en.htm (accessed 20 September).

European Commission (2002) (online) http://europa.eu.int/eurlex/en/lit/dat/1999/en_399D1419.html (accessed 28 February).

Evans, G. (1995) 'Tourism versus education – core functions of museums?', in Leslie, D. (ed.) *Tourism and Leisure: Culture, Heritage and Participation*, Brighton: LSA.

Evans, G. (1996) 'The Millennium Festival and urban regeneration – planning, politics and the party', in Robinson, M. *et al.* (eds) *Managing Cultural Resources for the Tourist*, Sunderland: Business Education Publishers, pp. 79–98.

Evans, G. (2001) *Cultural Planning: An Urban Renaissance?*, London: Routledge.

Falconer, C. (1999) '"The Dome does good" says Lord Falconer', *Cabinet Office Press Release*, CAB 301/99.

Featherstone, M. (1991) *Consumer Culture and Postmodernism*, London, and Newbury Park, CA: Sage Publications.

Featherstone, M. and Lash, S. (eds) (1999) *Spaces of Culture: City, Nation, World*, London: Sage Publications Ltd.

Feifer, M. (1985) *Going Places: The Ways of the Tourist from Imperial Rome to the Present Day*, London: Macmillan.

Feilden, B.M. and Jokilehto, J. (1998) *Management Guidelines for World Cultural Heritage Sites*, Rome: ICCROM.

Fernandes, C. and Sousa, L. (1999) 'Initiatives for developing textile crafts in the Alto Minho, Portugal', in Richards, G. (ed.) *Developing and Marketing Crafts Tourism*, Tilburg: ATLAS, pp. 55–72.

Financial Times (2001a) (online) *Portugal – Discovered Again: The EU Stimulus*, http://www.conway.com/portugal/9808/pg02.htm (accessed 20 September).

Financial Times (2001b) (online) *Economic Effect: Doubts over Games' Lasting Effects*, http://specials.ft.com/ln/ftsurveys/country/sc236ca.htm (accessed 21 Sepember).

Fisher, M. and Owen, U. (eds) (1991) *Whose Cities?*, London: Penguin.

Fourmile, H. (1994) 'Aboriginal arts in relation to multiculturalism', in Gunew, S. and Rizvi, F. (eds) *Culture, Difference and the Arts*, St Leonards: Allen & Unwin, pp. 69–85.

Fowler, P.J. (1992) *The Past in Contemporary Society: Then, Now*, London: Routledge.

Fox-Przeworski, J., Goddard, J. and De Jong, M. (1991) *Urban Regeneration in Changing Economy: An International Perspective*, Oxford: Clarendon Press.

Frideres, J.S. (1988) *Native Peoples in Canada: Contemporary Conflicts*, Scarborough, Ontario: Prentice Hall Canada.

Friedman, J. (1999) 'The hybridization of roots and the abhorrence of the Bush', in Featherstone, M. and Lash, S. (eds) *Spaces of Culture: City, Nation, World*, London: Sage Publications, pp. 230–256.

Gaarder, J. (2000) *Maya*, London: Phoenix.

Gale, R. (ed.) (1989) *Resources of Hope – Raymond Williams*, London: Verso.

Gambia Tourism Concern (2002) (online) http://www.gambiatourismconcern.com/ and http://www.subuk.net/tourism/ (accessed 15 September).

Garland, A. (1996) *The Beach*, London: Penguin.

Gathercole, P. and Lowenthal, D. (eds) (1994) *The Politics of the Past*, London: Routledge.

Getz, D. (1994) 'Event tourism and the authenticity dilemma', in Theobald, W.F. (ed.) *Global Tourism*, Oxford: Butterworth Heinemann, pp. 409–427.

Getz, D. (1997) *Event Management and Event Tourism*, New York: Cognizant Communication Corp.

Glass, R. (1964) 'Introduction: aspects of change', in Centre for Urban Studies, *London: Aspects of Change,* London: MacGibbon and Kee, pp. xiii–xlii.

Goffman, E. (1959) *The Presentation of Self in Everyday Life*, New York: Doubleday.

Gourevitch, P. (1999) 'Nightmare on 15th Street', *Guardian*, 4th December, pp. 1–3.

Government of the Northwest Territories (GNWT) (1983) *Community Based Tourism. A Strategy for the Northwest Territories Tourism Industry*, Yellowknife: Department of Economic Development and Tourism.

Graburn, N.H. (ed.) (1976) *Ethnic and Tourist Arts: Cultural Expressions from the Fourth World*, Berkeley: University of California Press.

Graburn, N.H. (2002) 'The ethnographic tourist', in Dann, G.M.S. (ed.) *The Tourist as a Metaphor of the Social World*, Wallingford: CABI, pp. 19–40.

Graham, B., Ashworth, G.J. and Tunbridge, J.E. (2000) *A Geography of Heritage: Power, Culture and Economy*, London: Arnold.

Grekin, J. and Milne, S. (1996) 'Towards sustainable tourism development: the case of Pond Inlet, NWT', in Butler, R. and Hinch, T. (eds) *Tourism and Indigenous Peoples*, London: International Thomson Business Press, pp. 76–106.

Gruffudd, P. (1995) 'Heritage as national identity: histories and prospects of the national pasts', in Herbert, D.T. (ed.) *Heritage, Tourism and Society*, London: Mansell Publishing, pp. 49–67.

Guggenheim Bilbao Museum (2001a) (online) http://www.bm30.es/proyectos/guggy_uk.html (accessed 20 September).

Guggenheim Bilbao Museum (2001b) (online) *Bilbao Thrives from the 'Guggenheim Effect'*, http://www.ulissul.com/mundo_bionico/Guggenheim.htm (accessed 20 September).

Gunew, S. and Rizvi, F. (eds) (1994) *Culture, Difference and the Arts*, St Leonards: Allen and Unwin.

Hadley, R. (1999) 'The arts and tourism: shot-gun wedding or long-term love affair?', *Arts-business*, 21 June, pp. 7–8.

Hajer, M.A. (1993) 'Rotterdam: re-designing the public domain', in Bianchini, F. and Parkinson, M. (eds) *Cultural Policy and Urban Regeneration: The West European Experience*, Manchester: Manchester University Press, pp. 48–72.

Hall, C.M. (1992a) *Hallmark Tourist Events: Impacts, Management, Planning*, London: Belhaven Press.

Hall, C.M. (1992b) 'Sex tourism in South-East Asia', in Harrison, D. (ed.) *Tourism in the Less Developed Countries*, London: John Wiley & Sons, pp. 64–74.

Hall, C.M. (1994) *Tourism and Politics: Policy, Power and Place*, Chichester: John Wiley & Sons.

Hall, C.M. (2000) *Tourism Planning: Policies, Processes and Relationships*, London: Prentice Hall.

Hall, P. (1989) *London 2001*, London: Unwin Hyman.

Hannerz, U. (1990) 'Cosmopolitans and locals in world culture', in Featherstone, M. (ed.) *Global Culture: Nationalism, Globalisation and Modernity*, London: Sage, pp. 237–252.

Harron, S. and Weiler, B. (1992) 'Ethnic tourism', in Weiler, B. and Hall, C.M. (eds) *Special Interest Tourism*, London: Belhaven Press, pp. 83–94.

Harvey, D. (1989) *The Urban Experience*, Oxford: Blackwell.

Harvey, D. (1990) *The Condition of Postmodernity*, Oxford: Blackwell.

Hatton, B. (2001) (online) *Lisbon Expo '98*, http://www.dispatch.co.za/1998/05/20/editoria/FAIR.HTM (accessed 20 September).

Hawkins, D.E. and Khan, M.M. (1994) 'Ecotourism opportunities for developing countries', in Theobald, W.F. (ed.) *Global Tourism: The Next Decade*, Oxford: Butterworth-Heinemann, pp. 191–204.

Herbert, D.T. (ed.) (1995a) *Heritage, Tourism and Society*, London: Mansell Publishing.

Herbert, D.T. (1995b) 'Heritage as literary place', in Herbert, D.T. (ed.) *Heritage, Tourism and Society*, London: Mansell Publishing, pp. 32–48.

Herder, J.G. ([1774] 1969) *Herder on Social and Political Culture*, ed. and trans. F.M. Barnard, Cambridge: Cambridge University Press.

Hewison, R. (1987) *The Heritage Industry – Britain in a Climate of Decline*, London: Methuen.

Hewison, R. (1991) 'Commerce and culture', in Corner, J. and Harvey, S. (eds) *Enterprise and Heritage: Crosscurrents of National Culture*, London: Routledge, pp. 162–177.

Hewison, R. (1997) *Culture and Cconsensus: England, Art and Politics since 1940*, London: Methuen.

Hinch, T. and Butler, R. (1996) 'Indigenous tourism: a common ground for discussion', in Butler, R. and Hinch, T. (eds) *Tourism and Indigenous Peoples*, London: International Thomson Business Press, pp. 3–19.

Hollinshead, K. (1997) 'Heritage tourism under post-modernity: truth and past', in Ryan, C. (ed.) *The Tourist Experience: A New Introduction.* London: Cassell, pp. 170–193.

Hope, C.A. and Klemm, M.S. (2001) 'Tourism in difficult areas revisited: the case of Bradford', *Tourism Management*, 22, pp. 629–635.

Horne, D. (1984) *The Great Museum*, London: Pluto Press.

Hotel Online (2001) (online) *Tourism and Convention Industries Impact is Profound for Olympic Host Cities,* http://www.hotel-online.com/Neo/News/PR2001_3rd/July01_OlympicCities.html (accessed 21 September).

Hoyau, P. (1988) 'Heritage and "the conserver society": the French case', in Lumley, R. (ed.) *The Museum Time Machine*, London: Routledge, pp. 27–35.

Hudson, K. (1996) 'Ecomuseums become more realistic', *Nordisk Museologi*, 2, pp. 11–20.

Hughes, G. (1998) 'Tourism and the semiological realization of space', in Ringer, G. (ed.) *Destinations: Cultural Landscapes of Tourism*, London: Routledge, pp. 17–33.

Hughes, H. (2000) *Arts, Entertainment and Tourism*, Oxford: Butterworth Heinemann.

ICOMOS (1999) *International Cultural Tourism Charter: Managing Tourism at Places of Heritage Significance*, 8th Draft, XIIth General Assembly.

Inskeep, E. (1994) *National and Regional Tourism Planning: Methodologies and Case Studies*, London: Routledge.

Inter-Commission Task Force on Indigenous Peoples (IUCN) (1993) *Indigenous Peoples and Sustainability: A Guide for Action*, IUCN Inter-Commission Task Force on Indigenous Peoples and the Secretariat of IUCN in Collaboration with the International Institute for Sustainable Development.

Ironbridge Gorge Museum (2001) (online) http://www.ironbridge.org.uk/ (accessed 15 June).

Irvine, A. (1999) *The Battle for the Millennium Dome*, London: Irvine News Agency.

Jacobs, J.M. (1961) *The Death and Life of Great American Cities*, New York: Penguin.

Jacobs, J.M. (1996) *Edge of Empire: Postcolonialism and the City*, London: Routledge.

Jamal, T.B. and Getz, D. (1995) 'Collaboration theory and community tourism planning', *Annals of Tourism Research*, 22, pp. 186–204.

Jamal, T.B. and Hill, S. (2002) 'The home and the world: (post)touristic spaces of (in)authenticity', in Dann, G.M.S. (ed.) *The Tourist as a Metaphor of the Social World*, Wallingford: CABI, pp. 77–108.

Jameson, F. (1984) 'Postmodernism, or the cultural logic of late capitalism', *New Left Review*, 146, pp. 52–92.

Jeng, M. (2002) *All-inclusive Holidays Will Deny us Even the Crumbs*, Gambia Tourism Concern (online) http://www.subuk.net/tourism/ (accessed 17 February).

Jermyn, H. and Desai, P. (2000) *Arts – What's in a Word?: Ethnic Minorities and the Arts*, London: Arts Council of England.

Jones, A.L. (1993) 'Contemporary issues in waterfront regeneration – a case study of the Swansea waterfront', M. Phil. thesis, University of Wales, Swansea.

Joneslanglasalle (2001) (online) http://www.joneslanglasalle.com.hk/Press/2001/160401.htm (accessed 21 September).

Jordan, G. and Weedon, C. (1995) *Cultural Politics: Class, Gender, Race and the Postmodern World*, Oxford: Blackwell.

Kalgeropoulou, H. (1996) 'Cultural tourism in Greece', in Richards, G. (ed.) *Cultural Tourism in Europe*, Wallingford: CABI, pp. 183–195.

Kavanagh, G. (ed.) (1996) *Making Histories in Museums*, London: Leicester University Press.

Keith, M. and Rogers, A. (eds) (1991) *Rhetoric and Reality in the Inner City*, London: Mansell Publishing.

Kempton, L. (2002) *Is All-inclusive Robbing our Visitors of Their Freedom?*, Gambia Tourism Concern (online) http://www.subuk.net/tourism/ (accessed 17 February).

Khan, N. (1976) *The Arts Britain Ignores: The Arts of Ethnic Minorities in Britain*, London: Commission for Racial Equality.

Kirschenblatt-Gimblett, B. (1998) *Destination Culture: Tourism, Museums and Heritage*, Berkeley: University of California Press.

Knox, P.L. (1993) 'Capital, material culture and socio-spatial differentiation', in Knox, P.L. (ed.) *The Restless Urban Landscape*, Englewood Cliffs, NJ: Prentice Hall, pp. 1–34.

Landry, C. and Bianchini, F. (1995) *The Creative City*, London: Demos Comedia.

Lanfant, M. (1995) 'Introduction', in Lanfant, M., Allcock, J.B. and Bruner, E.M. *International Tourism: Identity and Change*, London: Sage.

Larkham, P. J. (1995) 'Heritage as planned and conserved', in Herbert, D.T. (ed.) *Heritage, Tourism and Society*, London: Mansell Publishing, pp. 85–116.

Lavrijsen, R. (ed.) (1993) *Cultural Diversity in the Arts: Art, Art Policies and the Facelift of Europe*, the Netherlands: Royal Tropical Institute.

Lencek, L. and Bosker, G. (1999) *The Beach: The History of Paradise on Earth*, London: Pimlico.

Lennon, J.J. and Foley, M. (1999) 'Interpretation of the unimaginable: the U.S. Holocaust Memorial Museum, Washington, D.C., and "dark tourism"', *Journal of Travel Research*, 38 (August), pp. 46–50.

Lewis, J. (1990) *Art, Culture and Enterprise*, London: Routledge.

Li, V. (2000) 'What's in a name? Questioning "globalisation"', *Cultural Critique*, 45 (spring), pp. 1–39.

Light, D. (1995) 'Heritage as informal education', in Herbert, D. (ed.) *Heritage, Tourism and Society*, London: Mansell Publishing, pp. 117–145.

Lippard, L. (1990) *Mixed Blessings: New Art in a Multicultural America*, New York: Pantheon Books.

Lisbon Expo '98 (2001) (online) http://www.autodesk.co.kr/completestory/0,,66268—146820,00.html (accessed 20 September).

Livingstone, K. (2001) (online) *Carnival Needs to be Made Safe, But the Great Street Festival Must Go On*, http://www.london.gov.uk/mayor/articles/2000/indsep5.htm (accessed 16 October).

Lodge, D. (1991) *Paradise News*, Harmonsworth: Penguin.

Loughlin, J. (1997) '"Europe of the regions" and the Federalization of Europe: fact or fiction?', in *Papers from the ECTARC Convocation*, 11 July, Llangollen, Wales.

Lowenthal, D. (1985) *The Past is a Foreign Country*, Cambridge: Cambridge University Press.

Lowenthal, D. (1998) *The Heritage Crusade and the Spoils of History*, Cambridge: Cambridge University Press.

Lumley, R. (1988) *The Museum Time Machine*, London: Routledge.

Lumley, R. (1994) 'The debate on heritage reviewed', in Miles, R. and Zavala, L. (eds) *Towards the Museum of the Future: New European Perspectives*, London: Routledge, pp. 57–69.

Lyotard, J-F. (1986) *Le Postmoderne explique aux enfants*, Paris: Gallimard.

Maasai Environmental Resource Coalition (2002) (online) http://www.maasaierc.org/ecotourism.htm (accessed 15 March).

MacCannell, D. (1976) *The Tourist: A New Theory of the Leisure Class*, New York: Schocken.

MacDonald, D. (1994) 'A theory of mass culture', in Storey, J. (ed.) *Cultural Theory and Popular Culture – A Reader*, London: Harvester Wheatsheaf, pp. 28–43.

McCabe, S. (2002) 'The tourist experience and everyday life', in Dann, G.M.S. (ed.) *The Tourist as a Metaphor of the Social World*, Wallingford: CABI, pp. 61–76.

McLaren, D. (1992) 'London as ecosystem', in Thornley, A. (ed.) *The Crisis of London*, London: Routledge, pp. 56–68.

McGuigan, J. (1996) *Culture and the Public Sphere*, London: Routledge.

Maiztegui-Onate, C. and Areito Bertolin, M.T. (1996) 'Cultural tourism in Spain', in Richards, G. (ed.) *Cultural Tourism in Europe*, Wallingford: CABI, pp. 267–281.

Mann, M. (2000) *The Community Tourism Guide*, London: Tourism Concern/Earthscan Publications.

Massey, D. (1994) *Space, Place and Gender*, Cambridge: Polity Press.

Mathieson, A. and Wall, G. (1992) *Tourism: Economic, Physical and Social Impacts*, Harlow: Longman.

May, S. (1999) *The Pocket Philosopher: A Handbook of Aphorisms*, London: Metro.

Melzer, A.M., Weinberger, J. and Zinman, M.R. (eds) (1999) *Democracy and the Arts*, Ithaca and London: Cornell University Press.
Merriman, N. (1991) *Beyond the Glass Case: The Past, the Heritage and the Public in Britain*, Leicester: Leicester University Press.
Middleton, M. (1987) *Man Made the Town*, London: The Bodley Head.
Miettinen, S. (1999) 'Crafts tourism in Lapland', in Richards, G. (ed.) *Developing and Marketing Crafts Tourism*, Tilburg: ATLAS, pp. 89–103.
Miles, M. (1997) *Arts, Space and the City: Public Arts and Urban Futures*, London: Routledge.
Milner, A. (1994) *Contemporary Cultural Theory: An Introduction*, London: UCL Press.
Minerbi, L. (1996) 'Hawaii', in Hall, C.M. and Page, S.J. (eds) *Tourism in the Pacific: Issues and Cases*, London: International Thomson Business Press, pp. 190–204.
Moore, K. (2002) 'The discursive tourist', in Dann, G.M.S. (ed.) *The Tourist as a Metaphor of the Social World*, Wallingford: CABI, pp. 41–60.
Mowforth, M. and Munt, I. (1998) *Tourism and Sustainability: New Tourism in the Third World*, London: Routledge.
Mowitt, J. (2001) 'In the wake of Eurocentrism: an introduction', *Cultural Critique*, 47 (winter), pp. 3–15.
Murphy, P. (1985) *Tourism: A Community Approach*, New York: Methuen.
Muthyala, J. (2001) 'Reworlding America: the globalization of American studies', *Cultural Critique*, 47 (winter), pp. 91–119.
Myerscough, J. (1988) *The Economic Contribution of the Arts and Tourism*, London: Policy Studies Institute.
Nash, D. (1977) 'Tourism as a form of imperialism', in Smith, V. (ed.) *Hosts and Guests: The Anthropology of Tourism*, Oxford: Blackwell, pp. 33–47.
Nash, D. (1989) 'Tourism as a form of imperialism', in Smith, V. (ed.) *Hosts and Guests: The Anthropology of Tourism*, Philadelphia: University of Pennsylvania Press, pp. 37–52.
Negus, K. (1997) 'The production of culture', in Du Gay, P. (ed.) *Production of Culture/Cultures of Production*, London: Sage, pp. 67–118.
Nonini, D.M. (1999) 'Race, land, nation: a(t)-tribute to Raymond Williams', *Cultural Critique*, 41 (winter), pp. 158–183.
Owusu, K. (1986) *The Struggles for the Black Arts in Britain*, London: Comedia.
Owusu, K. and Ross, J. (1987) *Behind the Masquerade: The Story of the Notting Hill Carnival*, London: Arts Media Group.
Paasi, A. (2001) 'Europe as a social process and discourse: considerations of place, boundaries and identity', *European Urban and Regional Studies*, 9 (1), pp. 7–28.
Pacione, M. (ed.) (1997) *Britain's Cities: Geographies of Division in Urban Britain*, London: Routledge.
Parkinson, M. and Judd, D. (1988) 'Urban revitalisation in America and in the UK – the politics of uneven development', in Parkinson, M., Foley, B. and Judd, D. (eds) *Regenerating the Cities: The UK Crisis and the US Experience*, Manchester: Manchester University Press, pp. 1–26.
Paxman, J. (1998) *The English: A Portrait of a People*, London: Michael Joseph.
Pfafflin, G.F. (1987) 'Concern for tourism: European perspective and response', *Annals of Tourism Research*, 14 (4), pp. 576–579.
Philips, D. (1999) 'Narrativised spaces: the functions of story in the theme park', in Crouch, D. (ed.) *Leisure/tourism Geographies: Practices and Geographical Knowledge*, London: Routledge, pp. 91–108.
Plog, S.C. (1974) 'Why destination areas rise and fall in popularity', *Cornell Hotel and Restaurant Quarterly*, 14 (4), pp. 55–58.
Pocock, D.C.D. (1987) 'Haworth: the experience of a literary place', in Mallory, W.E. and Simpson-Housley, P. (eds) *Geography and Literature*, Syracuse, NY: Syracuse University Press, pp. 135–142.
Pocock, D.C.D. (1992) 'Catherine Cookson country: tourist expectation and experience', *Geography*, 77, pp. 236–243.

Pocock, D.C.D. (1997a) 'Some reflections on world heritage', *Area*, 29 (3), pp. 260–268.

Pocock, D.C.D. (1997b) 'The UK world heritage', *Geography: This Changing World*, 357, (82), pp. 380–384.

Porter, G. (1988) 'Putting your house in order: representations of women and domestic life', in Lumley, R. (ed.) *The Museum Time Machine*, London: Routledge, pp. 102–127.

Poulot, D. (1994) 'Identity as self-discovery: the ecomusuem in France', in Sherman, D.J. and Rogoff, I. (eds) *Museum Culture: Histories, Discourses, Spectacles*, Minneapolis: University of Minnesota Press, pp. 66–84.

Power, K. (1997) 'The material of change: Aboriginal cultures as a source of empowerment', in Landry, C. (ed.) *The Art of Regeneration*, London: Demos Comedia, pp. 52–56.

Punter, J. (1992) 'Classic carbuncles and mean streets: contemporary urban design and architecture in Central London', in Thornley, A. (ed.) *The Crisis of London*, London: Routledge, pp. 69–89.

Reed, M.G. (1997) 'Power relations and community-based tourism planning', *Annals of Tourism Research*, 24 (3), pp. 566–591.

Richards, G. (ed.) (1996) *Cultural Tourism in Europe*, Wallingford: CABI.

Richards, G. (1999a) 'Cultural capital or cultural capitals?', in Nystrom, L. (ed.) *City and Culture: Cultural Processes and Urban Sustainability*, Kalmar: The Swedish Urban Environment Council, pp. 403–414.

Richards, G. (1999b) 'Culture, crafts and tourism: a vital partnership', in Richards, G. (ed.) *Developing and Marketing Crafts Tourism*, Tilburg: ATLAS, pp. 11–35.

Richards, G. (2001a) 'The development of cultural tourism in Europe', in Richards, G. (ed.) *Cultural Attractions and European Tourism*, Wallingford: CABI, pp. 3–29.

Richards, G. (2001b) 'Cultural tourists or a culture of tourism? The European cultural tourism market', in Butcher, J. (ed.) *Innovations in Cultural Tourism*, Proceedings of the 5th ATLAS International Conference, Rethymnon, Crete, 1998, Tilburg: ATLAS,

Richards, G. (2001c) *Creative Tourism as a Factor in Destination Development.* ATLAS 10th Anniversary International Conference papers, 4–6 October, Dublin.

Richez, G. (1996) 'Sustaining local cultural identity: social unrest and tourism in Corsica', in Priestley, G.K., Edwards, J.A. and Coccossis, H. (eds) *Sustainable Tourism? European Experiences*, Wallingford: CABI, pp. 176–188.

Riley, R., Baker, D. and Van Doren, C.S. (1998) 'Movie induced tourism', *Annals of Tourism Research*, 25, (4), pp. 919–935.

Ritzer, G. (1993) *The McDonaldization of Society*, Thousand Oaks, CA: Pine Oak Press.

Ritzer, G. and Liska, A. (1997) '"McDisneyization" and "post-tourism": contemporary perspectives on contemporary tourism', in Rojek, C. and Urry, J. (eds) *Touring Cultures: Transformations of Travel and Theory*, London: Routledge, pp. 96–109.

Ritzer, G. and Liska, A. (2000) 'Postmodernism and tourism', in Beynon, J. and Dunkerley, D. (eds) *Globalization: The Reader*, London: The Athlone Press, pp. 152–155.

Robins, K. (1993) 'Prisoners of the city – whatever could a postmodern city be?', in Carter, E., Donald, J. and Squires, J. (eds) *Space and Place – Theories of Identity and Location*, London: Lawrence & Wishart, pp. 303–330.

Robins, K. (1997) 'What in the world's going on?', in Du Gay, P. (ed.) *Production of Culture/Cultures of Production*, London: Sage, pp. 11–66.

Rockwell, J. (1999) 'Serious music', in Melzer, A.M., Weinberger, J. and Zinman, M.R. (eds) *Democracy and the Arts*, Ithaca and London: Cornell University Press, pp. 92–102.

Roiter, F. (1991) *Venetian Carnival*, Venice: Zerella.

Rojek, C. (1993) *Ways of Escape: Modern Transformations in Leisure and Travel*, London: Macmillan.

Rojek, C. (1997) 'Indexing, dragging and the social construction of tourist sights', in Rojek, C. and Urry, J. (eds) *Touring Cultures: Transformations of Travel and Theory*, London: Routledge, pp. 52–74.

Rojek, C. and Urry, J. (eds) (1997) *Touring Cultures: Transformations of Travel and Theory*, London: Routledge.

Rolfe, H. (1992) *Arts Festivals in the UK*, London: Policy Studies Institute.

Ryan, C. (1991) *Recreational Tourism: A Social Science Perspective*, London: International Thomson Business Press.

Ryan, C. (1996) 'Some dimensions of Maori involvement in tourism', in Selwyn, T. (ed.) *The Tourist Image: Myths and Myth-making in Tourism*, Chichester: John Wiley & Sons, pp. 229–243.

Ryan, C. (2000) 'Sex tourism: paradigms of confusion', in Carter, S. and Clift, S. (eds) *Tourism and Sex: Culture, Commerce and Coercion*, London: Cassell, pp. 35–71.

Said, E.W. (1978) *Orientalism*, London: Routledge & Kegan Paul.

Said, E.W. (1993) *Culture and Imperialism*, London: Chatto & Windus.

Sampson, M. (1986) 'The origins of the Trinidad Carnival', in Arts Council of Great Britain, *Masquerading: The Art of the Notting Hill Carnival*, London: ACGB, pp. 30–34.

Sanchez Taylor, J. (1998) 'Embodied commodities', in *Embodied Commodities: Sex and Tourism*, Tourism Concern, *In Focus*, 30, (winter), pp. 9–11.

Sanchez Taylor, J. and O'Connell Davidson, J. (1998) 'Doing the hustle', in *Embodied Commodities: Sex and Tourism*, Tourism Concern, *In Focus*, 30 (winter), pp. 7–8, 17.

Sardar, Z. and Wynn Davies, M. (2002) *Why Do People Hate America?*, London: Icon Books.

Sarup, M. (1996) *Identity, Culture and the Postmodern World*, Edinburgh: Edinburgh University Press.

Schadler, F. (1979) 'African arts and crafts in a world of changing values', in De Kadt, E. (ed.) *Tourism – Passport to Development?*, New York: Oxford University Press, pp. 146–156.

Schouten, F.F.J. (1995) 'Heritage as historical reality', in Herbert, D. (ed.) *Heritage, Tourism and Society*, London: Mansell Publishing, pp. 21–31.

Seaton, A.V. (2002) 'Tourism as metempsychosis and metensomatosis: the personae of eternal recurrence', in Dann, G.M.S. (ed.) *The Tourist as a Metaphor of the Social World*, Wallingford: CABI, pp. 135–168.

Selwyn, T. (ed.) (1996) *The Tourist Image: Myths and Myth-making in Tourism*, Chichester: John Wiley & Sons.

Shackley, M. (ed.) (1998) *Visitor Management: Case Studies from World Heritage Sites*, London: Butterworth-Heinemann.

Sharpley, R. (1994) *Tourism, Tourists and Society*, Huntingdon: ELM Publications.

Simpson, M. (1996) *Making Representations: Museums in the Post-Colonial Era*, London: Routledge.

Smith, C. (1998) *Creative Britain*, London: Faber and Faber.

Smith, C. and Jenner, P. (1998) 'The impact of festivals and special events on tourism', *Travel and Tourism Intelligence*, 4, pp. 73–91.

Smith, K.A. (2000) 'The road to world heritage site designation: Derwent Valley Mills, a work in progress', in Robinson, M. *et al.* (eds) *Tourism and Heritage Relationships: Global, National and Local Perspectives*, Sunderland: Business Education Publishers, pp. 397–416.

Smith, M.K. and Smith, K.A. (2000) 'Surviving the millennium experience: the future of urban regeneration in Greenwich', in Robinson, M. (ed.) *Developments in Urban and Rural Tourism*, Sunderland: Business Education Publishers, pp. 251–267.

Smith, N. (1991) 'Mapping the gentrification frontier', in Keith, M. and Rogers, A. (eds) *Rhetoric and Reality in the Inner City*, London: Mansell Publishing.

Smith, N. (1996) *The New Urban Frontier: Gentrification and the Revanchist City*, London: Routledge.

Smith, V.L. (ed.) (1989) *Hosts and Guests: An Anthropology of Tourism*, Philadelphia: University of Pennsylvania Press.

Smith, V.L. (1997) 'The four *H*s of tribal tourism: Acoma – A Pueblo case study', in Cooper, C. and Wanhill, S. (eds) *Tourism Development: Environmental and Community Issues*, London: John Wiley & Sons, pp. 141–151.

Smith, V.L. and Eadington, W.R. (eds) (1994) *Tourism Alternatives: Potentials and Problems in the Development of Tourism*, Chichester: John Wiley & Sons.

Solomon, R.C. and Higgins, K.M. (1996) *A Short History of Philosophy*, Oxford: Oxford University Press.

Sorenson, C. (1989) 'Theme parks and time machines', in Vergo, P. (ed.) *The New Museology*, London: Reaktion Books, pp. 60–73.

Squire, S.J. (1993) 'Valuing countryside: reflections on Beatrix Potter tourism', *Area*, 24, pp. 5–10.

Standard Bank National Arts Festival (2001) (online) http://www.mg.co.za/mg/saarts/test-grahamstown1.htm (accessed 25 October).

Steele-Prohaska, S. (1996) 'The greatest story never told: Native American initiatives into cultural heritage tourism', in Robinson, M. *et al.* (eds) *Tourism and Cultural Change*, Sunderland: Business Education Publishers, pp. 171–182.

Strinati, D. (1994) 'Postmodernism and popular culture', in Storey, J. (ed.) *Cultural Theory and Popular Culture – A Reader*, London: Harvester Wheatsheaf, pp. 428–438.

Strubell, M. (1997) 'Regional autonomy: case study – Spain', in *Papers from the ECTARC Convocation*, 11 July, Llangollen, Wales.

Sudjic, D. (1993) *The 100 Mile City*, London: Flamingo.

Sudjic, D. (1999) 'Between the metropolitan and the provincial', in Nystrom, L. (ed.) *City and Culture: Cultural Processes and Urban Sustainability*, Kalmar: The Swedish Urban Environment Council, pp. 178–185.

Swarbrooke, J. (2000) 'Museums: theme parks of the third Millennium?', in Robinson, M. *et al.* (eds) *Tourism and Heritage Relationships: Global, National and Local Perspectives*, Sunderland: Business Education Publishers, pp. 417–431.

Sydney Gay Mardi Gras (2001) (online) http://www.mardigras.com.au/About/Historytable.html (accessed 12 February 2002).

Sydney Olympics 2000 (2001a) (online) http://www.olympicwebsite.com/venues.htm (accessed 21 September).

Sydney Olympics 2000 (2001b) (online) http://www.cpnonline.com/news/2001/070901_a.asp (accessed 21 September).

Tahana, N. and Oppermann, M. (1998) 'Maori cultural performances and tourism', *Tourism Recreation Research*, 23(1), pp. 23–30.

Taj Mahal (2001) (online) http://www.worldmonuments.org/html/progress/tajmahaprog.html (accessed 26 September).

Taylor, I., Evans, K. and Fraser, P. (1996) *A Tale of Two Cities: Global Change, Local Feeling and Everyday Life in the North of England: A Study in Manchester and Sheffield*, London: Routledge.

Te Papa Tongarewa Museum (2001) (online) http://www.tepapa.govt.nz/ (accessed 12 February).

Theobald, W.F. (ed.) (1994) *Global Tourism*, Oxford: Butterworth-Heinemann.

Tighe, A.J. (1986) 'The arts/tourism partnership', *Journal of Travel Research*, 24 (3), pp. 2–9.

Tilden, F. (1977) *Interpreting Our Heritage*, Chapel Hill: University of North Carolina Press.

Trask, Haunani-Kay (2001) (online) *Tourism and the Prostitution of Hawaiian Culture*, http://www.cs.org/newdirection/voices/rask5.htm (accessed 06 September).

Trask, M. (1998) 'Culture vultures', in *Indigenous Peoples, Human Rights and Tourism*, Tourism Concern, *In Focus*, 29 (autumn), pp. 14–17.

Tresidder, R. (1999) 'Tourism and sacred landscapes', in Crouch, D. (ed.) *Leisure/Tourism Geographies: Practices and Geographical Knowledge*, London: Routledge, pp. 137–148.

Trevor-Roper, H. (1965) *The Rise of Christian Europe*, London: Thames & Hudson.

Tunbridge, J.E. and Ashworth, G.J. (1996) *Dissonant Heritage: The Management of the Past as a Resource in Conflict*, London: John Wiley & Sons.

Turner, G. (1992) 'Tourism and the arts: let's work together', *Insights*, 3 (3), pp. A109–116.

Turner, L. and Ash, J. (1975) *The Golden Hordes: International Tourism and the Pleasure Periphery*, London: Constable.

UNESCO (1972) *Convention Concerning the Protection of the World Cultural and Natural Heritage*, Paris: UNESCO.

UNESCO (1982) *Culture Industries: A Challenge for the Future of Culture*, Paris: UNESCO.

U.S. Holocaust Memorial Museum (2001) (online) http://www.ushmm.org/ (accessed 15 January).

Urry, J. (1990) *The Tourist Gaze: Leisure and Travel in Contemporary Societies*, London: Sage.
Urry, J. (1999) 'Sensing leisure spaces', in Crouch, D. (ed.) *Leisure/Tourism Geographies: Practices and Geographical Knowledge*, London: Routledge, pp. 34–45.
Urry, J. (2002) *The Tourist Gaze*, London: Sage, 2nd edn.
Uzzell, D.L. (ed.) (1989) *Heritage Interpretation: Vol 1: The Natural and Built Environment*, London: Belhaven Press.
Van der Borg, J. and Costa, P. (1996) 'Cultural tourism in Italy', in Richards, G. (ed.) *Cultural Tourism In Europe*, Wallingford: CABI, pp. 215–231.
Varlow, S. (1995) 'Tourism and the arts: the relationship matures', *Insights*, 6 (4), pp. A93–98.
Von Eckardt, W. (1980) 'Synopsis', in *The Arts and City Planning*, Washington, DC: American Council for the Arts, pp. 136–142.
Wall, G. (1996) 'Towards the involvement of indigenous peoples in the management of heritage sites', in Robinson, M. *et al.* (eds) *Tourism and Cultural Change*, Sunderland: Business Education Publishers, pp. 311–320.
Walsh, K. (1992) *The Representation of the Past: Museums and Heritage in the Post-modern World*, London: Routledge.
Wang, N. (2000) *Tourism and Modernity: A Sociological Analysis*, Oxford: Pergamon Press.
Warren, S. (1999) 'Cultural contestation at Disneyland Paris', in Crouch, D. (ed.) *Leisure/Tourism Geographies: Practices and Geographical Knowledge*, London: Routledge, pp. 109–136.
Waters, M. (1995) *Globalization*, London: Routledge.
Weaver, D.B. (1998) *Ecotourism in the Less Developed World*, Wallingford: CABI.
West, B. (1988) 'The making of the English working past: a critical view of the Ironbridge Gorge Museum', in Lumley, R. (ed.) *The Museum Time Machine*, London: Routledge, pp. 36–62.
Wheatley, G. (1997) *World Heritage Sites*, London: English Heritage.
Whittaker, E. (2000) 'A century of indigenous images: the world according to the tourist postcard', in Robinson, M. *et al.* (eds) *Expressions of Culture, Identity and Meaning in Tourism*, Sunderland: Business Education Publishers, pp. 423–437.
Williams, R. (1958) 'Culture is ordinary', in Gale, R. (ed.) (1989) *Resources of Hope – Raymond Williams*, London: Verso.
Williams, R. (1976) *Keywords*, London: Fontana.
WOMAD (2001) (online) http://www.womad.org/ (accessed 24 October).
Wood, P. (1986) 'Economic change', in Clout, H. and Wood, P. (eds) *London: Problems of Change*, London: Longman, pp. 60–74.
World Heritage Committee (1994) *Convention Concerning the Protection of the World Cultural and Natural Heritage*, Paris: UNESCO.
Worpole, K. (1991) 'Trading places: the city workshop', in Fisher, M. and Owen, U. (eds) *Whose Cities?*, London: Penguin, pp. 142–152.
Worpole, K. (1992) *Towns for People: Transforming Urban Life*, Buckingham: Open University Press.
WTO (1993) *Recommendations on Tourism Statistics*, Madrid: WTO.
Wynne, D. (1992) (ed.) *The Culture Industry: The Arts in Urban Regeneration*, Aldershot: Avebury.
Zeppel, H. (1997a) 'Entertainers to entrepreneurs: Iban management of longhouse tourism in Sarawak, Borneo', in *Indigenous Cultures in an Interconnected World*, Proceedings of Fulbright Symposium Conference, Darwin, Australia, pp. 145–157.
Zeppel, H. (1997b) 'Maori tourism in New Zealand', *Tourism Management*, 18 (7), pp. 475–478.
Zeppel, H. and Hall, C.M. (1991) 'Selling art and history: cultural heritage and tourism', *Journal of Tourism Studies*, 2 (1), pp. 29–45.
Zeppel, H. and Hall, C.M. (1992) 'Arts and heritage tourism', in Weiler, B. and Hall, C.M. (eds) *Special Interest Tourism*, London: Belhaven Press, pp. 47–65.
Zukin, S. (1988) *Loft Living – Culture and Capital in Urban Change*, London: Century Hutchinson.
Zukin, S. (1995) *The Cultures of Cities*, Oxford: Blackwell.

# Index